Christian Climate Action is a community of Christians committed to prayerful direct action and public witness in response to climate breakdown. Inspired by Jesus Christ, and social justice movements of the past, CCA carries out acts of non-violent protest to urge those in power to make the changes needed.

For more information, visit the CCA website:

The general editor and spokesperson for this project is Jeremy Williams. Jeremy is an independent writer and campaigner specializing in communicating social and environmental issues to a mainstream audience. As well as his work with Extinction Rebellion, he has worked on projects for Oxfam, RSPB, WWF, Tearfund and many others. He is a co-founder of the Postgrowth Institute and co-author of *The Economics of Arrival: Ideas for a grown up economy* (Policy Press, 2019). His award winning website, The Earthbound Report, was recognized as Britain's number one green blog in 2018.

For more information, visit Jeremy's blog at <www.earthbound.report/about/>.

'The climate crisis is the biggest issue facing humanity today and it is unjust that those least responsible for causing it are facing its full effects. The scale of the emergency facing our world demands a just response from every one of us. I am delighted to see this important coming together of people of faith passionate about engaging their heads, hearts and hands in meeting this crisis. It is only together that we can make a difference.'
Amanda Khozi Mukwashi, CEO of Christian Aid

'Christians are called by God to show to the world what the divine image looks like — the image of a divine creator who brought the world to birth, called it good, and summoned human beings to reflect this divine care and delight through their own work in the world, animated by the gift of Christ's Spirit. This timely, moving and highly motivating book will help Christians of all ages to respond faithfully to that summons and grow more fully into the joyful responsibility we are made for.'
Rowan Williams, Master of Magdalen College, Oxford

'You'd think that those who commune daily with the Creator of the universe would be most aghast by the devastation of the creation, but that hasn't always been the case. Sometimes Christians have invalidated environmental activism because our theology has been exclusively concerned with going to heaven rather than restoring the earth. Here is some really good news. This is a landmark book in the movement of Christians who care deeply about the earth. It is nothing short of an invitation to join the holy uprising of people sweeping the globe who will not be silent in the face of the destruction of God's earth.'
Shane Claiborne, author, activist, and leader of Red Letter Christians

A resource book by the Christians in
Extinction Rebellion

TIME TO ACT.

Edited by
**Jeremy
Williams**

**Christian
Climate
Action**

spck

First published in Great Britain in 2020

Society for Promoting Christian Knowledge
36 Causton Street
London SW1P 4ST
www.spck.org.uk

Unless stated otherwise, Bible references are taken from the New
International Version, Biblica, 2011.

British Library Cataloguing-in-Publication Data
A catalogue record for this book is available from the British Library

ISBN 978-0-281-08446-3
eBook ISBN 978-0-281-08447-0

1 3 5 7 9 10 8 6 4 2

Typeset by The Book Guild Ltd, Leicester
First printed in Great Britain by Jellyfish Print Solutions

eBook by The Book Guild Ltd, Leicester

Produced on paper from sustainable forests

FSC
www.fsc.org

MIX
Paper from
responsible sources
FSC® C013604

'The growing possibility
of our destroying ourselves
and the world with our
own neglect and excess is
tragic and very real.

We cannot simply mourn
the fate of the earth.

We must do what we can.'

Billy Graham

Contents

Contents

Part 2
THE HEART

Contents

Contents

Part 3
THE HANDS

Acknowledgements

Thank you to all the activists and rebels who took the time to talk to us during the October Rebellion and on the phone, or who answered questions on the CCA website. This is your book. It would not exist without your willingness to share your stories and perspectives, hopes and fears.

With the book produced in a very short time, I owe a big thanks to all of those who made space in their diaries to write a chapter. It has been a privilege to get to read your words first, whether they made me cry or punch the air. I'd like to say a special thanks to those who do not normally consider themselves writers, and who took a step out of their comfort zone to contribute.

I am grateful to all those Christian Climate Action members who served as a sounding board for ideas, especially Ruth Jarman and Holly-Anna Petersen, and Caroline Harmon for the editing.

Thank you to all those at SPCK for seeing the potential for the book, and then pulling out all the stops to make it happen in record time.

Thank you to William Skeaping for his advice, and Clive and Charlie at This Ain't Rock-n-Roll for design and branding guidance. Several staff members at Tearfund were very helpful early on in the process, refining the proposal, clarifying the book's purpose and suggesting potential contributors. Hat tip to Jack Wakefield for naming the book.

Acknowledgements

Thank you to Louise, who helped with interviewing, transcribing, editing and moral support. And to Zach and Eden for joining in the rebellion. As Zach wrote on the day we got back from Blackfriars Bridge: "The world is changing. The climate's changing. And we are changing – standing on the bridge."

Introduction

JEREMY WILLIAMS

It's been a warm day for November, but it is cooling now as afternoon slips into evening. The wind is picking up on Blackfriars Bridge, snapping at the coloured flags of the rebellion. The street is beginning to clear after several hours of occupation. Looking down the Thames to the next bridge along, the double decker buses are on the move again.

As the crowd thins after the biggest act of civil disobedience in Britain for decades, the only people left are those who intend to stay. They are the activists who plan on getting arrested, who are locked together in the middle of the road. They will spend the night in a police cell to highlight the seriousness of the climate crisis, risking their freedom for the lives of others. And they are all Christians.

This is Extinction Rebellion, a mass movement calling for urgent action in the face of climate breakdown. Here at the sharp end of that movement is a small group of Christians who organise under the name Christian Climate Action. They are the last ones on this particular bridge, surrounded now by a ring of police officers working to separate them and arrest them, one by one. They are singing.

Like many others, I'm finding it hard to tear myself away. Some of the other activists are singing back to the Christians, from the other side of the police line.

I shouldn't be here at this point. I'm here with my children. It's time to get them home after a long day in the city, and the police are anxious to re-open the road. But this is beautiful. I don't want to miss it, and I'm not the only one who thinks so.

As we hover at the edges of the group, I notice a man with tears running down his cheeks. I edge closer to make sure he's okay, but I see he is with friends. Then I overhear him. "I grew up in the church," he is saying. "I walked away from it a long time ago, and since I left this is the thing that has spoken to me most. I don't know what I believe any more, but what's happening here is the most Christlike thing I've ever seen."

This book starts with Blackfriars Bridge partly because that's where it starts for me. I was outside the Houses of Parliament on the day that Extinction Rebellion launched, stepping into the road and sitting down to block it. But it was on the bridge a few weeks later that I understood the role of Christian Climate Action, a movement within a movement. It was here that I felt the Spirit of God at work, and an active curiosity about Extinction Rebellion became a commitment to give it real time and energy.

There is something symbolic about bridges too. They are neither one side nor the other. They are not a destination in themselves. Protest movements are the same. They are to get us from where we are to where we need to be, a means and not an end. It's a role the prophet Ezekiel describes, looking for someone who would 'stand in the gap' on behalf of the land, and prevent its destruction (Ezekiel 22.20).

More practically, we start here because this is where the

book comes from. This is not a book about activists, but *by* activists. It comes from the occupied streets and bridges and public squares of the rebellion. Literally so, as many of the voices included in the book were captured in interviews on the street during the actions of October 2019.

Finally, the book opens with the colourful disorder of a bridge occupation because this book reflects a real moment in time. It has been compiled, in haste, at a moment where we are learning and growing as a movement, being stretched in ways that are uncomfortable and even painful. There are so many unresolved questions, and it cannot hope to be comprehensive. It reflects the movement as it is, not what it will become. Like the crowd on the bridge, there is diversity in these pages – but not as much as there should be. There are flaws and compromises. This is not the last word.

What you will find in this book is a collection of short chapters on Christian non-violent direct action for the climate. There are essays, interviews, stories and poetry. It's a collection to dip in and out of, to inspire, challenge and motivate. It divides into three sections, and you can browse them however you like:

- **The Head** is all about thinking through the present moment, and it includes theological reflection, ethical argument, and learning from the legacy of Christian civil disobedience.
- **The Heart** deals with the experience and emotions of the climate emergency. There are lots of personal stories, and chapters on grief, burnout, and balancing action and rest.

- **The Hands** contains practical advice, and a wealth of resources to help you to find your place in the movement, and equip you for action.

It is my prayer that as you read these stories and thoughts, that you will be prompted to think about what this moment calls you to – this unique, chaotic, hopeful and terrifying moment that we find ourselves in. What will our response to climate breakdown be? Where will we stand on this issue of justice? What would Jesus do?

It is uncomfortable. It is difficult. I'm scared too.

But we are not alone. Our God is with us.

For ourselves then, and for each other. For our children, our grandchildren, and those as yet unborn. For the world's most vulnerable. For this gift of creation, and all our fellow creatures. For life itself, and for love.

Come and stand on a bridge with us.

Jeremy Williams is a writer and activist, a member of Christian Climate Action, and a blogger at <www.earthbound.report>.

Extinction Rebellion occupies
Blackfriars Bridge, November 2018.

THE
HEAD

1

We were not prepared for candles and prayers

RUTH VALERIO

On 9 October 1989, after a blessing by the bishop and an urgent call for non-violence, more than 2,000 people walked out of the St Nicholas Church Leipzig in what was then East Germany. They were met both by uniformed police who had been attempting to close the church for many months, and by tens of thousands of others, waiting outside with candles in their hands. It has been said that two hands are needed for carrying a candle: one to carry it and one to protect it from going out, and in this way one cannot also carry stones or clubs at the same time.

This was a pivotal moment in the fall of the Berlin Wall and the collapse of the dictatorship that had caused so many deaths and so much repression, brutality and misery.

During the 1980s, peace prayer services became a regular part of the life of the St Nicholas Church, beginning with only a handful of people, but growing into a fundamental part of the movement against the dictatorship. The church became a place where people could gather to discuss the urgent social problems and pray to God for support and guidance. As the movement grew, so it became infiltrated by Communist Party members who then heard Jesus' teachings from the Beatitudes and the Sermon on the Mount week by week. In fact, 600 of the people in the church listening to the bishop on 9th October were Stasi members.

Just two days earlier there had been a hideous show of armed force by the police as hundreds of defenceless people were beaten and taken away, but still people flocked to the Church on the 10th October, determined to continue their prayerful stand. The result was a peaceful movement that brought down the dictatorship: the military and the police became engaged in conversations and withdrew and there was not a single shattered window. Horst Sindermann, a member of the Central Committee of the ruling party said, "We had planned everything. We were prepared for everything. But not for candles and prayers."

Peaceful civil disobedience has a long history in the practice of the Christian faith. As I write in *Just Living: faith and community in an age of consumerism*, we sometimes forget that the earliest followers of the resurrected Jesus were themselves a subversive, minority group who refused to acknowledge anyone as Lord except Jesus Christ, and they suffered the consequences for doing so. The roots go back into the Jewish faith and particularly during times when the people of Israel found themselves in situations of exile or foreign oppression. When Mordecai warned Esther that thousands of lives were at risk at the hand of the authorities, Esther decided she had to speak up, even if that included peacefully breaking the law. She said: "Go, gather together all the Jews who are in Susa, and fast for me. Do not eat or drink for three days, night or day. I and my attendants will fast as you do. When this is done, I will go to the king, *even though it is against the law.* And if I perish, I perish" (Esther 4.16).

Jesus himself was clear that the law was there to serve people and the whole creation, and not the other way round. If an ox has fallen into a pit or a well on the Sabbath day, of course the owner would pull it out and not leave it there till the Sabbath is over – that is part of what it means to 'do good' (Matt 12.11; Luke 14.5). Jesus readily breaks the law and heals a man with a withered hand on the Sabbath saying, "I ask you, which is lawful on the Sabbath: to do good or to do evil, to save life or to destroy it?" (Luke 6.1–9)

So what about the situation that we find ourselves in today? In October 2019, the second wave of XR protests took place and it was my privilege to stand on the 'Faith Bridge' at Lambeth Bridge. Tearfund joined with CAFOD, Christian Aid, Christian Climate Action and others in an act of prayer and worship, praying for climate justice for people living in poverty, and declaring our commitment to care for God's whole creation.

It was a poignant moment for me because just that weekend I had received a message from Philip, who lives in the central part of Tanzania. I met Philip a few years ago when he presented me with a cockerel in a box (which unfortunately I couldn't take home with me, though the people I was staying with were pleased with the gift and I believe it may have graced their table not long after). I have since helped him with his chicken business. Not long after, his father sadly died and Philip has been left looking after the family. Philip has a younger brother studying at medical school, but Tanzania is suffering the impact of climate breakdown and the rains have not come. Because the rains have not come, food has not grown which means there are food shortages and the price for food has rocketed up. Food is so expensive that Philip now cannot afford to pay the bills for his brother to do his second year at medical school and Philip wanted to ask if I might help.

The climate crisis is leading to a food crisis which is leading to an economic crisis which is leading to a social crisis, all of which results in people like Philip suffering.

I have now been speaking and writing about the terrible problems of our changing climate for over twenty years – and others of course have been doing so for longer than that. I looked back at some old notes recently and realized with a sickening thud that the things I was talking about as predictions back then have now become reality. Over the years my language has changed from 'this will happen' to 'this is happening now'. I dug out an old powerpoint and saw there was a quote from a leading UN climate scientist who said we had until 2012 to make the change: if we did not change by then it would be too late.

Yet here we are, some years past 2012, and we still have not changed: our global emissions continue to rise and we have stayed firmly wedded to fossil fuels for way too long. We know that we face disaster: we are already seeing increasingly extreme weather events; sea level rise; species extinction and melting glaciers. From a human perspective, health, livelihoods, food security, water supply, human security and economic growth are all seriously at risk. Yet the higher our emissions go, the more terrible the impacts will be: the consequences of a 2°C warmer world are far greater than that of 1.5°C – though the impacts of the latter are still severe.

A recent UN report gives us ten years to change course, and says that the speed at which we do or do not cut our emissions between now and 2030 will make or break our prospects of keeping temperature rise to 1.5°. The world is way off track from reaching net zero even as late as 2050, and it is increasingly clear that even that goal is too late for millions of people. We need to have reduced our carbon emissions by 45 per cent in the next ten years. If we do not do this, it is likely we will see 100 million people being pushed back into poverty; global crop yield losses of as much as 5 per cent over the next ten years, and the loss of all our coral reefs.

As a follower of Jesus, I want to see people freed from poverty, living transformed lives and reaching their God-given potential. And I want to see a natural world that is flourishing, healthy and teeming with life; one in which we take our place as part of the whole community of creation. I do see signs of hope. Despite the huge frustrations, I do see governments and businesses starting to shift, and I see the Church, both in the UK and around the world, starting to wake up and take action. The Anglican Church of Southern Africa (comprising of some 3–4 million Anglicans and 2,000 clergy) has declared a climate emergency and committed to becoming a zero waste Church, and dioceses and other denominations across the UK are following suit, as are local councils.

Compared to the context I have been speaking into for most of the last twenty years or so, the wind is shifting and change is coming. But it is late and we will only change the current trajectory we are on and keep our world to within 1.5°C of warming if we take urgent action. That means moving from the incremental changes that have been happening, to deep and transformational change. I want to acknowledge that the change is happening, not because of people like me working within the system for years, but because of those who have recognized the situation is so desperate that only acts that disrupt the system will make the difference needed.

It is because of all of this that, while Tearfund itself does not endorse illegal activity, it felt absolutely right to stand with the protesters on the Faith Bridge, as Christians, to play our part in the call that is coming from around the world for deep and ambitious action on our climate crisis, and to add our voices and prayers. Personally speaking, I would not have wanted to have been anywhere else. From the start of the rebellions in April 2019, I knew I needed to get on a train and be there.

We worship the Lord of all creation, and our primary allegiance is to him – to a God who calls us to act justly, love mercy and walk humbly (Mic 6.8). We face a situation where we do not see justice being done, where we do not see mercy being shown to those living in poverty or to other creatures, and where we see leaders of both nations and corporations proudly acting in their own economic interests. In the words of Dietrich Bonhoeffer regarding resistance to the Nazi state, the role of the church in such times is 'not just to bandage the victims under the wheel, but to put a spoke in the wheel itself'.

Environmentalist and theologian, social activist and author, Dr Ruth Valerio is Global Advocacy and Influencing Director at Tearfund and Canon Theologian at Rochester Cathedral.

'The climate is a
common good,
belonging to all
and meant for all.'

Pope Francis

2

The story of Christian Climate Action

CAROLINE HARMON

Since Autumn 2018 Christian Climate Action has mostly joined in with Extinction Rebellion protests, but we have been around since 2012.

Many of our founding members had been taking action on climate change for many years before they got involved in Christian Climate Action (CCA). We had signed petitions, contacted and met with our MPs, made changes big and small to our lifestyles, brought up the issue in our church communities, been members of organizations such as Green Christian, A Rocha and Operation Noah, and encouraged others to get involved. Yet we all recognized that this just wasn't achieving what needs to be achieved anywhere fast enough. Some of our members had already taken Non-Violent Direct Action (NVDA) around other issues and recognized the power of NVDA to bring about change.

Early on we created this statement to explain, both to ourselves and to others, what we are about:

Christian Climate Action is a community of Christians supporting each other to take meaningful action in the face of imminent and catastrophic anthropogenic climate change. Inspired by Jesus Christ, and social justice movements of the past, we carry out acts of non-violent direct action to urge those in power to make the change needed.

In 2013 some of our members attended a Reclaim the Power camp in Balcombe, Sussex, which aimed to stop fracking in the area. Two of our members were arrested alongside many others, including Caroline Lucas MP. They went to court charged with breach of a section 14 order and wilful obstruction of the highway and were found not guilty.

In February 2015 we decided to take action because the Church of England wasn't divesting from fossil fuels. A request to discuss this at Synod had been turned down, so we wrote what we wanted to say on a large banner and two of our members entered the public gallery. They dropped the banner over the side of the gallery so that everyone could read it. It said:

We are young Christians, for us and our children climate change is the biggest threat we face. We ask you to pray and act on behalf of all those afflicted by climate change now and in the future.

As a church community, we cannot continue to invest in fossil fuel companies. So we ask you, on our behalf, to divest now.

'May God defend the afflicted among the people and save the children of the needy' (Psalm 72.4).

At the same time, several of our members held a vigil outside the building. Word had got out to climate change campaigners, both Christian and not, and we had a gathering of around 20 people (probably our largest ever at the time)! We sang, prayed, read scripture, talked about why we were there and lit candles.

At the end of 2015, on the first day of the Paris Conference (COP21), which led to the Paris Agreement, five members of CCA whitewashed the walls of the Department for Energy and Climate Change (as it was then called). At the time the government claimed to be the 'greenest government ever', but we felt this was whitewash and we wanted to make that clear. CCA members prayed and worshipped outside the building while the five waited to be arrested.

The following spring they were in court charged with criminal damage. All five were found guilty. A number of our members held a vigil all day outside the building and we took it in turns to enter the public galley in the court to support those on trial. Afterwards we set up our first crowdfunder to pay their fines and raised the money we needed within days.

We organized actions at this building again on the first day of Advent in 2016 and in 2017, which included delivering melting ice to the building, but no one was arrested on these occasions.

Undeterred by the possibility of a criminal record, in May 2016, five of our members (three of whom had been on trial just the month before) helped shut down the UK's largest open-cast coal mine at Ffos-y-Fran for one day. The action was organized by Reclaim the Power and around 400 people took part.

Over the years we've also used street theatre several times to protest. In May 2017 we staged a wedding on the steps of Church House in Westminster between the Church and the fossil fuel industry. We did this to highlight the Church of England's refusal at the time to divest from the fossil fuel industry because they preferred a policy of 'engagement'. The play had a happy ending, with the Bride of Christ leaving Mr Fossil Fuels at the altar and running off with Jesus!

We've also supported anti-fracking campaigns, visiting Preston New Road during 'No Faith in Fracking' week and holding two one-day prayer vigil tours of the East Midlands to visit various sites associated with proposals for fracking.

In 2018 we were approached by Extinction Rebellion to see if we would like to join forces with them. We loved their well thought out, non-violent approach and began to work with them. Some of our members were involved in early swarming protests, which tested the ground for the larger Rebellions that have happened since.

Then we got involved in the October 2018 Rebellion, the Easter 2019 Rebellion and the October 2019 Rebellion. For the October 2019 Rebellion we were invited to help create a Faith Bridge – a multi-faith location devoted to prayer and reflection above the Thames. While the bridge itself was reclaimed by the police on the day, the week saw a whole series of actions alongside other XR faith groups, and solidarity actions from leading Christian NGOs.

Our involvement in XR has transformed CCA in many ways. In October 2018 we had maybe 40 people on our mailing list and our actions would attract 5–20 members. Today we have close to 900 people on our mailing list and we estimate that more than 100 CCA members attended the October 2019 Rebellion. Often during our actions one person used to be at home sending out a press release at the appropriate moment and taking press enquiries. Until recently this involved maybe doing one or two interviews with Christian media outlets. During the Easter 2019 Rebellion we suddenly found our phones ringing off the hook with media requests from both Christian and secular media, and our email box was overflowing with questions about how people could join us. We struggled to keep up for a while! Sometimes people are shocked that we don't have a great deal of infrastructure behind us or, until very recently, any paid staff (and we still only have one paid person for one day a week).

During 2019 we ran several national training days for new members and dozens of our members started providing talks and sermons for churches and organizations around the UK (we now get requests each week for such talks). We also now have regional groups across the UK and groups in countries from The Netherlands to Australia acting under the name 'Christian Climate Action'.

At our heart we're still just a bunch of ordinary people taking action around our jobs and responsibilities, to care for God's creation as best we can. We'd love for you to join us on this crazy journey – the best part about it is meeting others who care so much about the climate emergency.

Get in touch at <www.christianclimateaction.org>.

Caroline Harmon did everything she could think of to care for creation and encourage others to do the same, before coming to the conclusion that prayer combined with direct action is our best hope. She has been involved with CCA since 2013.

3

The third way – what Jesus said about non-violence

INSPIRED BY WALTER WINK

When thinking of the non-violence movements that have inspired Extinction Rebellion, Mohandas Gandhi is one of the first names that comes to mind. He was in turn shaped by many others, including Jesus. Gandhi read the New Testament for the first time as a student in London, and he wrote in his autobiography that the "Sermon on the Mount went straight to my heart." One particular section, Matthew 5.38–41, was to become foundational to his teaching:

> You have heard that it was said, 'Eye for eye, and tooth for tooth.' But I tell you, do not resist an evil person. If anyone slaps you on the right cheek, turn to them the other cheek also. And if anyone wants to sue you and take your shirt, hand over your coat as well. If anyone forces you to go one mile, go with them two miles.

The subtitle that modern Bible translators use for this passage give us some clues about how it has been interpreted. The New Century Version goes for 'Don't fight back', and the New Living Translation captions it 'Teaching about revenge'. For others, including Leo Tolstoy, Dorothy Day or Gandhi, the passage was about resistance.

'Turning the other cheek' is not about submission, or passively accepting bullying and intimidation. Christians are not called to be doormats. There is something else going on in this passage, and one of the writers who has explained it most clearly is the theologian and activist Walter Wink. "The gospel does not teach nonresistance to evil," he clarifies in his book *The Powers That Be*. "Jesus counsels resistance, but without violence."

Part of the problem is our understanding of 'resist' in Jesus' line "do not resist". The Greek word here is *antistenai*, to 'stand against'. As Wink points out, this is a military term that describes the meeting of two armies. With this definition, what Jesus means is that we should not oppose with force. It's a rejection of violence, but not a call to be passive in the face of evil:

> Jesus is not telling us to submit to evil, but to refuse to oppose it on its own terms. We are not to let the opponent dictate the methods of our opposition. He is urging us to transcend both passivity and violence by finding a third way, one that is at once assertive and yet nonviolent.

Demanding equality

Having called his followers to this third way, Jesus gives them three practical applications as examples. Reading them in the 21st century, it is easy to miss the political context of these examples. For his original hearers, there would have been no mistake – Jesus was advocating creative resistance.

First, Jesus suggests turning the other cheek, but he's quite specific about which one – the right cheek. This matters, says Walter Wink. If you hit someone with your fist or your open palm, you will hit their left cheek. A slap to the right cheek is a back-hand, "not a blow to injure, but to insult, humiliate, degrade. It was not administered to an equal, but to an inferior." It's a strike that established dominance and hierarchy.

Turning and inviting a slap to the other cheek is to refuse to acknowledge inferiority. It insists on equality, directly challenging the master/slave relationship. It's entirely peaceful, but shockingly defiant. It's assertive, turning an attempt to shame and degrade back on the powerful, exposing and rejecting their claim to superiority.

Shaming the powerful

In his second application, Jesus talks to those who are being sued. For an audience of first century peasants, this kind of forced debt repayment would be familiar. Wealth flowed upwards from the poorest, to the Jewish elites, and to Rome at the top. The system could take everything from you, even the clothes off your back.

Here, again, a little local context is required. Most people only had two items of clothing – an outer and an inner garment. (This is what Jesus wore, as we see when the soldiers divide up his clothes in John 19.) Jesus suggests that if you are being sued for one of them, hand over both. That would of course leave you naked.

If you're being sued for your clothes, then you have nothing left. The powerful have taken everything else already, and now they've come for the pittance that remains – the final indignity. By stripping naked in the courtroom, the poor could expose the shamefulness of such an unjust and callous system. "The poor man has transcended this attempt to humiliate him," says Wink. "He has risen above shame. At the same time, he has registered a stunning protest against the system that created his debt."

There is hope for the creditor too in this symbolic stripping. "This unmasking is not simply punitive, since it offers the creditor a chance to see, perhaps for the first time in his life, what his practices cause, and to repent."

Dilemma situations

Finally, Jesus describes the common practice where Roman soldiers on the march were permitted to commandeer someone to carry their pack for a mile. That was their right. However, they had no right to request a second mile. That was breaking the rules.

Understood that way, you can see the problem that Jesus sets up for the Roman soldier. Having had a break from carrying his pack for a mile, he pauses to take it back – only for his press-ganged porter to refuse. A mile back down the road it was the soldier who was in charge. Now, embarrassed, he has to ask for his pack back so that he doesn't get into trouble.

"From a situation of servile impressment," says Wink, "the oppressed have once more seized the initiative. They have taken back the power of choice. They have thrown the soldier off balance by depriving him of the predictability of his victim's response. He has never dealt with such a problem before. Now he must make a decision for which nothing in his previous experience has prepared him. If he has enjoyed feeling superior to the vanquished, he will not enjoy it today."

Offering to go a second mile with the soldier isn't an act of kindness. It's an act of subversion. It challenges oppressive power by asserting the agency of the oppressed.

Both sides win

In the old dichotomy of fight or flight, one has to choose between attacking or running away. If we run, the system goes unchallenged, and our own safety is the most we can hope for. If we fight, we are no better than the oppressor – and likely to lose anyway. Fight or flight can only deliver one winner, and it is likely to be the most powerful.

Jesus proposed a third way, and crucially, this third option opens up the possibility for transformative change. Both sides can win: the oppressed take back their power, and a mirror is held up to the oppressor. They are confronted with the truth, and that creates an opportunity to repent. "Jesus is not advocating nonviolence merely as a technique for outwitting the enemy, but as a just means of opposing the enemy in a way that holds open the possibility of the enemy's becoming just also. Both sides must win. We are summoned to pray for our enemies' transformation, and to respond to ill treatment with a love that is not only godly but also from God."

Third way non-violence believes in the humanity of the oppressor too. That is what it means to love your enemies, which is the very next thing Jesus tells the crowd. "Love your enemies and pray for those who persecute you, that you may be children of your Father in heaven."

Applying the third way

The examples that Jesus gives are very specific to his time and his audience. We are unlikely to be forced to carry a soldier's pack for them. In Britain at least, there would be legal consequences for a boss who routinely slapped their employees. The challenge for us is to look for the principles in play in Jesus' suggestions, and work out how to use them in our context of non-violent direct action for the climate.

In his book on non-violence in South Africa, Walter Wink provides a summary. Between submission and withdrawal on the one side (flight), and armed rebellion and revenge on the other (fight), is Jesus' third way.

Jesus' Third Way
- Seize the moral initiative
- Find a creative alternative to violence
- Assert your own humanity and dignity as a person
- Meet force with ridicule or humour
- Break the cycle of humiliation
- Refuse to submit or to accept the inferior position

- Expose the injustice of the system
- Take control of the power dynamic
- Shame the oppressor into repentance
- Stand your ground
- Make the Powers make decisions for which they are not prepared
- Recognize your own power
- Be willing to suffer rather than retaliate
- Force the oppressor to see you in a new light
- Deprive the oppressor of a situation where a show of force is effective
- Be willing to undergo the penalty of breaking unjust laws
- Die to fear of the old order and its rules

During the October Rebellion in London in 2019, police were wrong-footed by an army of nursing mothers. Arriving with babies in slings and pushchairs, they occupied the street, sat down and began to breastfeed. Here were activists – dismissed as anarchists and trouble-makers in the press – caring for their children with the most simple and natural act of nurturing. Here were the children that will inherit a broken climate, carried by their mothers who refuse to stand by and let that injustice happen. And here were the police, flicking through their training manuals in vain for advice on arresting breast-feeding mothers. I think it would make Jesus smile.

4

The ethics of civil disobedience

SIMON KITTLE

When considering the ethical status of an action, an important first step is to seek clarity on the nature of the action, and the intentions behind it. To that end, it is worth beginning by reflecting on the nature of civil disobedience.

Civil disobedience is a *considered, public, nonviolent breach of the law*. It typically aims to highlight an unjust aspect of society in order to bring about positive change. Examples *may* (depending on the laws of the society in question) include mass demonstrations and boycotts, refusing to pay taxes, blocking or occupying buildings, roads, and private or public spaces, striking from work or school and other forms of non-cooperation with the state.

Civil disobedience is *considered* in two ways. First, thought is given to the inconvenience and disruption that will be caused. Second, those engaging in it do not attempt to escape notice, but instead stand ready to give an account of their actions and why they are thought to be justified (in contrast to the criminal who attempts to avoid being caught). A powerful example of this occurred on the first day of the Paris Climate Change Conference in November 2015, when a group from Christian Climate Action performed the symbolic action of whitewashing the walls of the UK's Department for Energy and Climate Change (DECC), before rebranding it the Department for Extreme Climate Change. Those involved handed a letter to the DECC beforehand to explain their actions and afterwards waited for the police to arrest them. They took responsibility for what they had done, arguing in court that they were not guilty given the DECC's lack of action on global heating.

Civil disobedience is usually public, since this helps to raise awareness of the injustice. It is nonviolent, not just because there is an absence of violence, but because there is an express intention to act peacefully and to avoid harm to others. It typically involves a breach of the law, because participants either (i) consider some particular law to be unjust, or (ii) believe that breaking a law is the only way to draw attention to a wider structural injustice.

Extinction Rebellion (XR) practices nonviolent civil disobedience "in full public light", with organizers and participants taking responsibility for their actions "in an attempt to halt mass extinction and minimize the risk of social collapse". Tactics include "economic disruption to shake the current political system and civil disruption to raise awareness". One of the goals is to bring about enough political system change to secure meaningful action on global heating. However, it does not mean advocating for one political party over another – 'political' here is used in the broad sense to mean *the organization and administration of human communities*. The goal is a democracy where every citizen has "real and equal agency", and where an elite with vested interests in the fossil fuel industry could not hide the truth about the climate emergency and block meaningful action on lowering emissions.

Yet is it morally permissible to participate in civil disobedience? In particular, is it a legitimate *response to government inaction on the climate emergency*? For Christians, a comprehensive answer to these questions would ideally be based in a political theology drawn from scripture, experience and prior theological thinking. This is a difficult task for several reasons, including (i) the many texts in the Bible which reflect complex attitudes to the ruling authorities of the day, (ii) the widely differing cultural landscapes, both across the Bible's timespan and compared with today, and (iii) the unique nature of global heating as an ethical problem. Yet even without a fully developed political theology, some progress on the ethics of civil disobedience can still be made.

Consider the opening verses of Romans 13: "Let everyone be subject to the governing authorities, for there is no authority except that which God has established … Consequently, whoever rebels against the authority is rebelling against what God has instituted" (NRSV). One traditional interpretation applies these verses more or less directly to Christians everywhere and everywhen, teaching that the ruling authorities are, to use the words of John Calvin, "constituted by God's ordination" to provide a "just government of the world" so that evil may be restrained. This thought can be found as far back as Irenaeus (c. 130– 202 AD). And since God has established the authorities for this purpose, everyone owes them "a common duty of obedience". Yet even in this tradition of deference to the state, thinkers have pointed out that if the ruling authorities were instituted by God to rule justly, some form of resistance is justified when they fail in this duty. In support, Peter's exclamation in Acts 5.29 is often cited: "We must obey God rather than human beings!" (NRSV)

An excellent case can be made for saying that, for Christians, this resistance or rebellion should be nonviolent, but the present point is just that sometimes breaking the law is the right thing to do. Civil disobedience cannot not be ruled out ahead of time. Whether it's permissible in any particular case will depend upon the details of that case. On reflection, this should not be surprising. Jesus's action in the Temple is arguably a classic case of symbolic civil disobedience. Paul wrote many of his letters from prison after being arrested for disturbing the peace, and tens of thousands of early Christians followed in these footsteps.

What, then, is the case for civil disobedience in the face of government inaction on global heating? While it is not possible here to discuss the ethical issues around specific actions, e.g. blocking a road vs. striking from school, two classes of fact provide a strong case for some kind of civil disobedience as a response to the climate emergency.

First, facts concerning the seriousness of global heating. At the Paris Climate Summit in 2015 over 190 countries agreed to keep global heating below 2°C above pre-industrial levels, and to "make efforts" to keep it below 1.5°C. Yet aside from a brief dip due to the financial crisis, and mostly stable emissions in 2015, the trend is rising emissions – and rising temperatures. More than 50 per cent of all industrial CO_2 emissions in the atmosphere today have been produced in the last 30 years. According to Climate Action Tracker, a baseline scenario of either no policies or non-implementation of policies will lead to a warming of at least 4.1°C. If all current policies were implemented, the result would be about 3.2°C warming; even an 'optimistic' policy estimate results in around 2.9°C of heating. The effects of this level of warming would be devastating: a 2014 World Bank report said that unchecked global warming could kill 250,000 people annually from 2030 onwards, and lead to 143 million climate refugees by 2050. Crop yields will fall, despite the world needing 50 per cent more food by 2050. Major world cities will be routinely inundated by floods. As David Attenborough concluded, the result of inaction could well be "the collapse of our civilizations".

Second, facts concerning the inaction of the authorities, taking recent UK governments as an example. In 2015, David Cameron's government created the UK Oil and Gas Authority with the stated mission of "[maximizing] the economic recovery of oil and gas", while also removing subsidies for onshore wind power in England. It ended the zero carbon homes initiative, increased VAT on solar panels, strongly encouraged the UK's fracking industry, and maintained 'tax relief' for North Sea oil and gas. In 2016 the UK government gave the fossil fuel industry the largest subsidies of any EU state, over 12 billion euros. One of Theresa May's first acts as Prime Minister was to dissolve the Department for Energy and Climate Change. Her government promoted fracking and airport expansion, while continuing to block onshore wind power. In 2018, May personally announced a freeze in fuel duty – the ninth consecutive yearly freeze. At the time of writing, Boris Johnson's government has only been in power a couple of months but David King, a leading scientist and British Ambassador to the Paris Climate talks, said he would give Johnson's government "3 or 4 out of 10" for their environmental stance. It is an open question whether the opposition would have done any better.

The two sets of facts sketched above provide a strong case for the legitimacy of civil disobedience as a response to the climate emergency. Further thought must always be given to the ethics of specific actions. Due to the nature of civil disobedience and the risk of arrest, each person considering it should be convinced in their own mind, not just that it is morally permissible, but that it is appropriate for them in their circumstances. Neither XR nor CCA should demand it of any of their members, for while it is morally *permissible*, that does not make it morally obligatory. Finally, it is important for those who, like the present author, have benefited from various forms of privilege, to approach the topic humbly, recognizing and seeking to learn from the many environmental activists across the world who have been engaged in civil disobedience at great risk to themselves for many decades.

Simon Kittle is a researcher in philosophy and theology based in Leeds.

5

An appeal to the church from a youth striker

FRANCESCA LAVEN

Driven by greed, adults have put the future of this planet and its inhabitants at risk. We are facing a climate emergency.

I took part in my first Youth Strike for Climate in March 2019, after seeing photos of the February strike in the news. I wanted to be part of the masses of students calling for change. I was tired of feeling isolated; I had enough of standing frozen before the injustices affecting the voiceless.

On the streets, I have seen the flame of anger in so many young people. We are tired of empty promises and petty politics. This is an emergency. This is beyond politics. All we want is a place to live, for all our fellow creatures and us. For too long we have been ignored. Now we refuse to be silenced.

On the streets, I have seen sparks of hope. The school strikes, along with other environmental groups, have drawn more attention to the climate emergency than ever before, and more adults have begun to listen. In November 2019, Pope Francis announced plans to add 'ecological sin' to the Catechism of the Catholic Church, a sacred duty to protect our common home.

Along with hope, the strikes have given purpose and community to me and the many young people who felt powerless in the face of this man-made catastrophe. We want the adults to take notice of and take action against the climate crisis. We acknowledge that many have been campaigning for years. Now, we ask all of you to lend your voices to ours. It is our future – your children and grandchildren's future – that is at stake; it is our burden to inherit.

Young people do not have money to spend. We have no platform from which to speak. We cannot vote. By refusing to attend school, we exert the only power we have.

Yet churches have power: moral, financial and political – from the bishops in the Lords, to each of you at the polling station. You can set an example for other institutions. You can encourage and celebrate your school strikers. Church buildings can provide sanctuary for them. You can speak up. You can take part in actions.

So support us. Join us. Make some noise. Let us use this power to defend our common home and our futures.

This is an emergency.

It is time to act.

Francesca Laven is 16, lives near Reading and has taken part in youth strikes for the climate.

6

The Faith Bridge – the importance of multi-faith actions

The climate emergency is 'beyond politics', as Extinction Rebellion says. It is bigger than party politics and requires us to work together across traditional boundaries. The same is true of religion. We all share one earth, and whatever differences and disagreements there may be between faiths, those can be put aside for shared action around the climate. More than that – the common cause of the climate can in fact build connections between faiths, creating shared spaces and opportunities to learn from each other.

"I've never been very engaged in multi-faith work for its own sake," says Rabbi Jeffrey Newman. "It's not been an area of special interest to me. What I've always felt is that working across faiths becomes vital and exciting when we have a project in common, when there's something that is beyond any of our particular abilities or vision. The climate change work is an extremely good example of that."

The most notable expression of this multi-faith approach was the Faith Bridge, planned as one of several sites in the October Rebellion. The vision was to occupy Lambeth Bridge, creating a tent village and decorating it with a garden, shrines and prayer flags, around the centrepiece of a large wooden ark. It would serve as a hub for a programme of worship, prayer, meditation and ritual.

Taking the bridge

The plan was to occupy the bridge on the first day of the rebellion and hold it for as long as possible. However, activists arrived on the Monday morning to find a heavy police presence. Activists from the faith groups were able to occupy one end of the bridge, and others from the South West of England held the other, with the police preventing anyone from crossing into the middle. Much of the infrastructure for the site had been intercepted on the way, and it looked as though the police would soon clear the area entirely.

Fortunately, Christian development charities had organized a solidarity march that morning, an entirely legal action in support of the Faith Bridge. With representatives from Tearfund, Christian Aid and CAFOD, they arrived at the bridge and the police stepped back to allow speeches to take place.

"I believe in a God of justice, a God of peace and a God of solidarity" Amanda Mukwashi, chief executive of Christian Aid, told the crowd. "I pray that we find courage, that we find strength, but we do so with humility, with a peaceful heart, and the wisdom that can only come from God above."

The NGOs unrolled a banner along the pavement, with a quote from William Wilberforce: 'You may choose to look the other way, but you can never again say that you did not know'.

Bringing the sides together

"I was there on the Faith Bridge on the first day," says Mothiur Rahman from XR Muslims. "We managed to hold the space while Christian Climate Action and CAFOD and others were speaking. We kept the schedule we had planned and did the call to prayer. In the evening the South West region that was holding the North side got some infrastructure down and we were invited to go and join them. The police wouldn't allow us to go across the bridge, so we had to go all the way round. We had a procession, all of us singing from the CCA songbook."

The Faith Bridge activists walked down the Thames path to Westminster Bridge, crossed over and came back up the other side. "Walking towards the mass protest on Westminster bridge singing Amazing Grace was definitely one of the highlights of the rebellion," says CCA's Ruth Jarman. "It was like the meeting of two armies in Lord of the Rings. I felt people understood and appreciated us people of faith."

Satya Robyn from XR Buddhists also names that walk as a highlight: "It was so moving to stand side-by-side with the Christians, and with Mothiur, and sing Amazing Grace. We took the road and walked down the bridge. We started to see other rebels down the road, and there was this moment of moving forwards, and really singing. It was about community coming together, and being united by a cause that takes priority over everything else."

The ark

The ark, built out of pallets by carpenter Rich Keal and a group of young people, had never made it to the bridge. The police had stopped the van and unloaded the ark onto the pavement. Then they repacked it to confiscate it, but couldn't get all of the pieces back in. They left behind the prow and the stern, the two largest sections. Activists picked them up and despite their weight, carried them almost a mile across Westminster Bridge to the other side.

"The only bit of infrastructure we had left on our side was Noah's Ark," says Rahman, "and that was being carried at the back. As we came towards the end of the walk they were getting tired, so I said I would help to carry it, and we came to the front. As the police saw us approaching they heard Amazing Grace and saw this wooden structure, and thought 'what is this?'"

"I stood and watched the police removing sections of the ark," said Gaynor Jenkins from Christian Climate Action. "It was so sad. Seeing it reappear, at least in part, was inspiring. Most importantly for me, it was a visual rallying point for us faith folk as we merged into a massive number of rebels on the North end of the bridge, with the police crushing us from behind. I admit I was terrified. But as I watched brave rebels sitting in the ark and surrounding it I smiled in the midst of my fear – a moment of God-given clarity."

The police wouldn't let the ark and the procession through, and a police liaison officer was sent for. While they were waiting, activists sat in the ark and others made speeches. "We are XR and we are the Ark," said one.

Mothiur Rahman also took the mic: "I started talking about the ark – 'we're carrying Noah's Ark here, and the ark was all about truth, and whose truth you're listening to. Nobody listened to Noah when he was trying to tell God's truth, and look what happened! Are we in a similar situation now where we've got climate change and nobody is listening to the scientists?' And the police arrested me then."

The ark proved to be the last stand for the Faith Bridge, with rebels remaining on it and around it as the police cleared the area. "I was arrested because of the ark," said another member of Christian Climate Action. "It was a huge symbol for me of faith, witness, life and death."

Regrouping in the square

The Faith Bridge itself was only held for a day, but the vision for a multi-faith site continued. Faith groups relocated to Trafalgar Square and pitched marquees between the lions, creating a courtyard of faith instead, with XR Jews, XR Buddhists, XR Muslims and Christian Climate Action. There were regular times of Franciscan prayer, Quaker worship or Buddhist meditation, with activists moving from one to the other. "It was beautiful for the CCA tent to serve the other faiths," says Ruth Jarman. "We hosted an Imam and Sufi singing. There was so much mutual respect between the people of different faiths."

"When there was a Muslim call to prayer, Christians held and respected that space," said one CCA member, and that did not go unnoticed. "Doing the call to prayer and having the group of Christians below supporting, I felt much more energized," says Rahman, who stood at the feet of one of Trafalgar Square's lions with a megaphone. "I felt welcomed into this space where the land has traditionally been Christian land."

"What I really liked is that different groups brought their own voices," says Satya Robyn. "There's a lot of noise, with the samba bands and the helicopters above. We did some walking meditations, in silence, and just witnessing. Holding the silence and holding the space felt important, and it's something we can bring to the movement." Then she adds, "Having said that, I'm a Buddhist who chants, so meditation isn't so much my thing. I really enjoyed the Christian hymns, really belting them out!"

"I was inspired by that sense of the presence of different faiths being together and taking part in something that is broader than any of our particular remits," said Rabbi Jeffrey Newman. CCA member David Jenkins also named the sense of unity between faiths in Trafalgar Square as a highlight of the Rebellion: "Walking around the square in love and grief with new Buddhist friends. Joining in four-part harmony with Jewish friends. Kissing the earth after the Muslim call to prayer." Jim Dandy echoes the sentiment, remembering "the unity of the faith bridge – Muslims, Jews, Buddhists and Christians all supporting one another in our acts of protest as prayer and worship." "Being part of a switched-on community of faith," said CCA member and Buddhist meditator Ruth Urbanowicz, "it was electric."

Taking action together

From the base at Trafalgar Square, activists from different faiths joined each other at actions. "I was walking up with the Faith group from the tube station to the Bank of England action," says Rachel from CCA. "I suddenly realized that I'd ended up in the middle of the XR Jews action, but felt strongly that I was in the right place. I went into 'clergy spouse mode', taking hold of prayer books, passing them back, relieving the guy leading prayers from the bags he was trying to juggle. Singing 'hoshana' as the police moved in to arrest Rabbi Jeffrey, then carrying all their stuff as they sang him to the police van."

Rahman from XR Muslims also supported Rabbi Newman. "The rabbi kneeled down in the road and it was very moving to see him. I read out his statement, and I think that's the service aspect. We don't just have to be talking about our thing. We can be of service to others, bringing their story forward, because all the stories have connections."

Phil Kingston also collaborated with XR Buddhists on a separate action. "I was asked by three Buddhists if I would accompany them in an action. I gladly responded, having seen the initiatives of recent popes in arranging inter-faith meetings in Assisi and having grown in understanding of the extent to which all of the great religions are committed to care for the Earth."

Satya Robyn spoke of her respect for those risking arrest. "I had this sense that there were people on the Faith Bridge who were led by their faith, and I think that's increasingly rare in our society, in our culture. There was something very strong about that, the experience of not just doing it for the planet but doing it for God, for humanity. That spiritual foundation is very empowering."

The witness of the Faith Bridge

As the police closed down protest across the city, the makeshift faith camp would also be eventually cleared. "It was a difficult night," says CCA's Ray Leonard, "helping to salvage what we could from the Faith Tent, and just seeing people's property being trashed was painful." Though the site was not set up again, activists did return in the morning. "Reclaiming the square peacefully and spiritually was a joy and very moving," says Leonard. "We placed the Faith banner in the centre, and formed two circles, and were led in singing. It was a moment for me that showed God cannot be denied, is known and moves within us all, and we have great hope and love."

A large banner with the words 'Faith Bridge' continued to make its way around protests throughout the rebellion, at one point leading a march through London, or serving as a central point for a candlelight vigil. It became a symbol of unity, a 'bridge' between faiths, and perhaps something more than that. As CCA's John Clements put it, "The rebellion really felt like being at the intersection of two worlds, the world that is and the world that could be."

XR JEWS

עושׂה שׁלום אֳפּני

Rabbi Jeffrey Newman kneels in
the road in the City of London.

7

Tell the truth – civil disobedience in the Jewish tradition

RABBI JEFFREY NEWMAN

I knew I would have to go through arrest. It wasn't something I was looking forward to or that I wanted. I just felt that until I had summoned the courage to do that, I was somehow not fully engaged.

The particular day on which it happened turned out to be the harvest festival of Sukkot, with its symbols of the fruits of the earth, and a day on which we were going to have a service. We came to a point in the road where the police had lined up on both sides. They wanted all the protesters to be on one side of the road, and the other side to be clear. They quickly tried to hustle us across, and I didn't feel that I particularly wanted to be hustled. So I stopped, and I read the Solemn Intention Statement. (See Chapter 39.) I started reading that out and then people echoed it. Then I read more loudly, and it became really strong.

To allow myself to even consider civil disobedience is to overturn a lifetime's submissiveness which some might suggest is a hallmark of the Jew. For 2,000 years, since our expulsion from Israel under the Romans, wherever we have lived in the world it has always been as a minority. We have been subjects of kings or religious authorities, paying the taxes demanded of us and remaining inconspicuous as far as possible. The few moments of exception have been rare.

True, since the emancipation allowed us to enter civil society, some individual Jews have been rebels in thought and action. But these few have seldom been consciously affected by Jewish teachings or traditions, still less have they attempted to act as leaders of the community.

Perhaps only in modern America, where despite being a tiny minority (1.5 per cent), Jews have felt secure enough that a few, perhaps a very few, have been prepared to protest government actions. Outstanding amongst these was Rabbi Abraham Joshua Heschel who marched to Alabama alongside Martin Luther King.

Arthur Waskow is another leading American rabbi. He has been arrested 26 times, from the sixties onwards, for protesting against racial segregation, the Vietnam and Iraq wars, Apartheid and the Soviet Union's oppression of Jews. My own thinking owes much both to his courage and his writings, particularly *The Sword and the Plowshare as Tools of Tikkun Olam*. Waskow highlights the astonishing success of the campaign in the Soviet Union by outstandingly courageous Jews taking non-violent direct action, within the country and outside, to stand against the regime. Though some were severely punished, eventually emigration, which had been prohibited, became possible. Freedom to identify and practise as Jews was gained. It may be that this Jewish resistance led eventually to the undermining of the oppressive state.

As Rabbi Waskow demonstrates, the very roots of Judaism are in revolution – out of a determined opposition to the prevailing culture, out of opposition to Pharaoh, even out of Abraham's preparedness to oppose God when, in the story of Sodom and Gomorrah, Abraham felt that God might act unjustly (Genesis 18). Indeed, without the preparedness of the two midwives, Shifra and Puah, to risk the wrath of Pharaoh and not kill Jewish boys at birth, Moses himself would never have survived (Exodus 1.15–18).

Judaism's insistent demand for social justice as shown by prophets such as Jeremiah, Isaiah or Elijah brought me from a completely secular upbringing to becoming a rabbi (which means teacher.) Of those, maybe Jeremiah is the prophet for our times. He lived at a time of upheaval, approximately 647–587 BCE. His humanity and acute political and religious understanding grasped that in the vast struggle for domination between Assyria, Babylon and Egypt, Judah and Jerusalem were mere pawns. Their only hope would be to remain outside the struggle, to take care of the needs of their own people, the poor, the widow, the orphan. He could see far beyond the superficial soothing words of the 'false' prophets, the greed of the wealthy and the self-interest of the kings.

While at college, I studied Jeremiah's terrible words about the future siege of Jerusalem which caused him to be put in the stocks (ch. 19 & 20). It didn't stop him. Released, he sits outside the king's house (Greta Thunberg outside the Swedish Parliament?) and shouts out warnings. A few years later when the exile to Babylon which he had foreseen happens, he once again warns the King not to attempt to

43

ally with Egypt and is thrown in jail. I admired Jeremiah's fortitude. Never did I think more than study might be necessary.

As I read the Solemn Intention Statement, the police were quiet and listening. We all thought that once it was done we would do what they wanted and cross the road. But what happened was that as I finished the declaration, there was no way that I was going to cross the road. I just wanted to sit down on the earth – even though it was tarmacked – and just be close to the earth. It wasn't an act of deliberate defiance. It was a necessity.

I'd already been warned by the inspector that I'd be arrested under Section 14 if I didn't move on. I said yes, that was fine and I was ready for it. Then he handed over to a police constable to actually make the arrest, because the inspector doesn't want to do the paperwork. He said, "I'm arresting you under Section 14", and he went on to say that I was obstructing the highway and police cars and ambulances couldn't get past. I said, "That's not true".

You can look at the footage of the arrest and see that I'm in the middle of the road and there's plenty of room to get past. "That's not true. You're not telling the truth." Of course, our number one demand is to tell the truth. He then said that anything I said could be taken down and used against me, and I replied that I wanted him to take it down. He didn't, and I said "You're not telling the truth, you're not taking it down! I'm going to make a complaint against you for not telling the truth."

It was one of those theatrical moments. It wasn't what I was aiming for, but because there were so many photographers there, it was captured in many different ways. I had on my Tallit (prayer shawl) and a Kippah with the XR logo, and I was holding the symbols of connection with the earth. That made it a powerfully symbolic religious moment, and that's part of what captured people's imagination. I've been very humbled by the impact my actions have had.

With the birth of Extinction Rebellion, we need to rebel to do all that we can to avoid a Sixth Mass Extinction and an unimaginable future for our children and our children's children. I must myself face the challenges of civil disobedience to arouse those who sleep, and help prepare us all for some likely unwelcome consequences of a sharp reduction of our carbon emissions. Now the issue of our times is not Brexit, not even jockeying for supremacy between superpowers, but the future well-being of life on the planet, humanity's responsibility for the planet itself. Now the words of Isaiah, for example chapter 24.19–20 (NKJV), seem especially close:

The earth is violently broken,
The earth is split open,
The earth is shaken exceedingly.
The earth shall reel to and fro like a drunkard,
And shall totter like a hut;
Its transgression shall be heavy upon it,
And it will fall, and not rise again.

When times such as these are at hand, who cannot rebel – rebel for life?

The rebellion must start with ourselves, with finding and sharing light – no blame, no identifying the faults of others. Only responsibility, or perhaps response-ability: the ability to respond to the great needs of our time.

Neither is the rebellion only negative. Another great Jewish teacher, Martin Buber, emphasized that the best way to bring about the future you desire is to begin the build it in miniature within the present. This must be the work of Extinction Rebellion. As Waskow writes: "No longer a passive non-violent protest against the world we disdain, Jewish nonviolence today must actively and positively create the world we want."

So may it be.

Rabbi Jeffrey Newman is Emeritus Rabbi of Finchley Reform Synagogue, and founder of Earth Charter UK.

8

Learning from Martin Luther King

In April 1963, Martin Luther King was part of a campaign of non-violent direct action in Birmingham, Alabama. It was the most segregated city in America at the time, and civil rights activists organized marches and sit-ins to disrupt the city and change the conversation.

Martin Luther King was arrested, declined to post bail, and wrote a famous letter from the Birmingham jail. It was written to local clergymen who had released a statement condemning the action. Running to several pages, his letter is one of the most influential documents in the history of non-violent direct action. It touches on many central principles for climate change activists today.

One of the first things King does is to emphasize the cause:

> You deplore the demonstrations taking place in Birmingham. But your statement, I am sorry to say, fails to express a similar concern for the conditions that brought about the demonstrations... It is unfortunate that demonstrations are taking place in Birmingham, but it is even more unfortunate that the city's white power structure left the Negro community with no alternative.

This is an important lesson for anyone who finds themselves in a conversation about Non-Violent Direct Action (NVDA). It is very easy to talk entirely about the legitimacy of the tactics, rather than the central issue of climate breakdown. Right at the start, King frames his letter around the core injustice.

Next, he contextualizes the actions for his sceptical readers. The actions are not off the cuff or anarchic, but part of a planned campaign with four steps.

- First, activists assess the facts and establish whether there is an injustice – a very obvious yes in the case of Birmingham.
- Second, activists attempt to negotiate. The city authorities had not cooperated, and business leaders had not kept their promises. Negotiation had therefore failed.
- Third, the activists prepared. King calls this 'self-purification', as did Gandhi, but it involved tools that would be familiar to Christian Climate Action members today: training in NVDA, workshops on de-escalation, and time spent in prayer.
- Finally, the fourth step is the direct action itself. It comes as a last resort, when there really is no viable alternative.

King's opponents argued that he should pursue negotiation instead of direct action, and King agreed:

You are quite right in calling for negotiation. Indeed, this is the very purpose of direct action. Nonviolent direct action seeks to create such a crisis and foster such a tension that a community which has constantly refused to negotiate is forced to confront the issue. It seeks so to dramatize the issue that it can no longer be ignored.

The creating of tension may seem at odds with non-violence, but King distinguished between a fearful, destructive tension, and "a constructive, non-violent tension which is necessary for growth." The direct action would disrupt, break through the status quo, and open the way for negotiation. In fact, King muses later on, the tension is already there. "We who engage in nonviolent direct action are not the creators of tension. We merely bring to the surface the hidden tension that is already alive. We bring it out in the open, where it can be seen and dealt with."

Extinction Rebellion echoes this strategy, using civil disobedience to raise awareness and crack the silence over the climate crisis. The 'creative tension' that results from direct action creates new possibilities. It resets people's expectations of what a proportionate response to climate change might be. It gives politicians permission to be more ambitious. It opens up new language. The civil rights movement exposed the tension in a racially segregated society. Today's climate activists drag unhealthy denial into the sunlight, exposing our inaction in the face of climate breakdown.

Like many who oppose direct action, the white clergy of Birmingham objected to the civil rights movement breaking laws. King acknowledged the concern, but argued that there were just and unjust laws. People of conscience have a legal and moral responsibility to disobey the unjust ones. As XR advocates today, King asked that those breaking an unjust law should do it "openly, lovingly, and with a willingness to accept the penalty."

Of course, the kinds of laws that the civil rights activists were breaking were very specific, and there was a direct connection between the action and the cause. They were sitting in cafés and diners where black customers were not served, or riding in bus seats reserved for white passengers. By breaking these laws, activists asserted their equal rights, empowered as individuals and collectively.

Climate change is so vast, the causes so distant from the effects, that it is harder to draw direct lines between the injustice and the protest action. We have to pick our targets wisely and work harder to explain ourselves, but the reality is that it is very ordinary things that are the problem – driving a car, flying on holiday, eating beef or running gas central heating. That is why Extinction Rebellion seeks to disrupt 'business as usual'. The breakdown of the climate is caused by such mundane things, and those with the highest carbon footprints are shielded from the damage. It is increasingly clear that the consumerist 'way of life' is ultimately a way of death, and that is what we rebel against.

To some, that makes us extremists. Martin Luther King faced the same accusations, which he resented at first, and then came to accept. "Was not Jesus an extremist for love?" he asks, demanding that his followers love their enemies and do good to those that hated them? Many figures down through the history of the church would be called extremists. "So the question is not whether we will be extremists, but what kind of extremists we will be" King concludes. "Will we be extremists for hate or for love? Will we be extremists for the preservation of injustice or for the extension of justice?"

Towards the end of his letter King confesses something that, as a pastor himself, obviously grieves him. He writes honestly about his disappointment with the church, and the lack of support from white ministers and congregations that he thought would rally behind the freedom movement. There are notable exceptions, but too many churches "have been more cautious than courageous and have remained silent behind the anesthetizing security of stained glass windows."

This is a sorrow that is familiar to many Christian climate activists. Environmental issues are often marginal. In some places, Christians have been slow to accept the science and recognize the moral implications of the climate crisis. We know that the church could be a powerful agent for change, but it can seem pre-occupied with doctrinal questions or personal morality. Many activists long to see the church wake up and engage. That is our hope and prayer, and many activists would recognize King's description of the tears he has shed over the church's inaction: "Be assured that my tears have been tears of love. There can be no deep disappointment where there is not deep love."

Reading King's letter today, there are many resonances with direct action for the climate, not least because there is a tragic continuity in the injustice. The climate crisis disproportionately affects people of colour. Black Lives

Matter UK highlighted this by obstructing London City Airport, three years before Christian Climate Action members were arrested there with Extinction Rebellion. There are significant differences between our context and that of a segregated America, but Martin Luther King's words continue to inspire, calling us to expose the tensions of climate change with integrity, compassion and prayer.

9

Re-engaging with the powers

VANESSA ELSTON

Wednesday of the first week of the October Rebellion, I am praying with a group of Christians in Trafalgar Square. We turn to the four points of the compass and voice our lament for the destruction of creation and tears stream down my face. Looking at the symbols of Empire in the square: the lions that surround Nelson's Column; the winged bull Assyrian deity that occupies the fourth plinth; the neo-classical columns of the National Gallery; I think about the intractable nature of what we are up against, something that feels more than just the sum of all the strands that make up 'business as usual'. The language of the Beast from the public reading of The Book of Revelation in Trafalgar Square later that afternoon feels strangely appropriate.

On the Monday of that week I had witnessed a friend from CCA being arrested on Lambeth Bridge. The act of arrest in a peaceful rebellion takes on a slightly surreal and almost ritualistic aspect. I watched him being politely warned by the police, he replying quietly why he won't move and then lying down in the road, relaxing the body to become softly non-compliant, so that it took five police men and women to lift and carry him to the van. We applauded as each person was arrested, voicing our love and respect for both arrestee and police. Just before my friend was arrested, I said to him "it is a form of prayer" and Romans 12.1 flashes across my mind: "Present your body as a living sacrifice". I am beginning to grasp that non-violent civil disobedience is essentially putting your body in the way and that there is more going on than just the inconvenience of blocking a bridge and using up police time.

The following Monday I am in the heady mix of protest, theatre and sociability that makes up the XR rebellion outside the Bank of England. It is definitely not 'business as usual'. I feel drawn to pray at the edge of the crowd, holding up the homemade sign I had quickly penned before rushing out the door: THE LOVE OF MONEY IS KILLING US – PRAY FOR A BETTER WAY. The XR Buddhists are meditating, and I gravitate to the edge of their space, holding vigil in the energy of their silence. I close my eyes and start to pray 'Come Lord, Come Holy Spirit. Have mercy on us, Forgive us, Change our hearts, Release our minds', and I practice, as best I can, the presence of God, in the heart of the finance district.

I wonder if I am imagining the smell of incense that fills the air. When I open my eyes I see that someone has actually placed two charcoal burners containing Frankincense at my feet. People are arranged differently in the space around me, it feels like something intangible is happening. I talk to people who would not align themselves with traditional forms of Christianity about spirituality and the need for deep change, about how "Jesus feels very relevant here". Someone from XR media asks if she can film me in silent prayer, with subtitles that capture what I am praying. She later emails me that Rabbi Jeffrey Newman, age 77, had been arrested there earlier that morning, kneeling on the street, a powerfully fragile figure in his prayer shawl.

These experiences not only awaken my inner activist but also question what it is that we are really up against in this "period of enormous, catastrophic breakdown" to quote the Rabbi. In 1982 the American Bible scholar Walter Wink went on sabbatical in Chile to complete a writing project "while experiencing life under military dictatorship". By the end of the trip he had become ill and overwhelmed by despair. "I had gone to Latin America hoping that what we experienced there would help me write a book that could make a difference. The evils we encountered were so monolithic, so massively supported by our own government, in some cases so anchored in a long history of tyranny, that it scarcely seemed that anything could make a difference". Unable to write the book he originally intended, Wink set out to discern the nature of structural evil through a systematic study of the language of the powers and principalities in the New Testament. Beginning with a "rather naïve assurance" that the powers could be understood solely as "human institutions, social systems and political structures", he found that there was always "this remainder, something that would not reduce to physical structures – something invisible, immaterial, spiritual and very, very real." Rather than depicting this 'something' as disembodied demons or hostile spirits, Wink argued that the "New Testament prefers to speak of the Powers only in their concretions, their structural inertia, their physical embodiments in history."

THE LOVE OF MONEY IS KILLING US - PRAY FOR A BETTER WAY

NON-VIOLENT

WAKE UP TO WHAT REALLY MATTERS

Vanessa Elston prays in the City of London.

"Every power tends to have a visible pole, an outer form – be it a church, a nation, or an economy – and an invisible pole, an inner spirit or driving force that animates, legitimates, and regulates its physical manifestation in the world. Neither pole is the cause of the other. Both come into existence together and cease to exist together."

Dietrich Bonhoeffer expressed a similar understanding about the invisible dimension of community in his earliest work exploring a theology of sociality: that "Where wills unite, a 'structure' is created – that is a third entity, previously unknown, independent of being willed or not willed by the persons uniting."

I want to suggest that we can apply this logic to the global corporations and financial networks that operate as the real power behind nation states today. William Dalrymple's latest book *The Anarchy* tells the story of the East India Company, from its unpromising beginnings in the City of London in 1599, to becoming the most powerful and violent for-profit corporation that has ever existed, responsible for the colonization of India in the eighteenth century. "Western imperialism and corporate capitalism were born at the same time, and both were to some extent the dragon's teeth that spawned the modern world." Dalrymple concludes that we need to heed the "ominous warning about the potential for the abuse of corporate power" in a world where "Empire is transforming itself into forms of global power that use campaign contributions and commercial lobbying, multinational finance systems and global markets, corporate influence and the predictive harvesting of the new surveillance-capitalism... to effect its ends."

Corporations have an inner driving force that goes beyond their employees and shareholders. They manifest a culture, ways of thinking, deeply embedded unspoken assumptions about 'the way things are and have to be.' Walter Brueggemann understood this 'spirit of Empire' as that which refuses to permit another imagination: "the ideology of our age does not believe in real newness… so it must defend, guard and protect at all cost the old." What Corporate Capitalism has yet to grasp is that the change needed requires more than a few modifications. What is required is a deeper change of mind, a new spirit that can only come through repentance, *metanoia*, a fundamental paradigm shift. As a society we are all caught up in a kind of group madness, we are in the grip of an idolatrous spell, hurtling towards the brink of destruction, the currents of global finance seemingly irresistible, all pervasive, yet strangely hidden and sustained in secrecy, unwilling to "come to the light, so that their deeds may be exposed." (John 3.20)

Many have rejected charismatic notions of spiritual warfare that, in Walter Wink's words, "Largely ignore the institutional sources of the demonic", and fail to do the "hard political and economic analysis to name, unmask and engage these Powers transformatively." However, I want to argue that we need not throw the baby out with the bath water, but need a more nuanced, non-reductive understanding of the nature of the Powers. We need new ways of grasping that the fight against climate breakdown is not primarily against "enemies of flesh and blood" (Ephesians 6.12). We are in a spiritual battle that requires spiritual weapons that "are not merely human", but "have divine power to destroy strongholds", and to "destroy arguments and every proud obstacle raised up against the knowledge of God." (2 Cor. 10.4–5 NRSV)

I am not speaking here of some kind crusade to impose our beliefs on others. There can be no return to Christendom. I am speaking of the need to reclaim an older wisdom in the language of spiritual warfare, of the Christian life as unavoidably engaging in the cosmic battle between good and evil that plays itself out in both the public and personal spheres of life. We may wince at military metaphors or the associations with physical violence, but there is a danger in the softer spirituality of 'the oneness of all', that we lose an understanding of the reality and urgency of the struggle "against the cosmic powers of this present darkness, the spiritual forces of evil in the heavenly places" (Ephes 6.12). We are called to "take up the whole armour of God", to take our stand for the meek of the earth, for those most vulnerable to climate and ecological breakdown, for all the creatures of God's good creation.

We are up against huge forces of inertia, self-interest and wilful blindness. We are up against the myths of modernity and progress. Where once Capitalism's self-justification was that it ensured individual freedom against totalitarian oppression, we are now fighting a bigger battle, more hidden and systemic. We have essentially been sold a lie: that the dignity of human liberty depends on our freedom to profit and consume without limit, oblivious to the physical limits of the created order on which we depend.

Wink wrote that "when a particular Power becomes idolatrous, placing itself above God's purposes for the good of the whole, then that Power becomes demonic. The church's task is to unmask this idolatry and recall the Powers to their created purposes in the world."

Wink's essential insight is to understand that as with human beings so with the Powers: "The powers are good, the powers are fallen, the powers must be redeemed." The church is not to roll over and acquiesce to our current system, rather we are to name and unmask the spell of limitless growth, the unquestioning faith in technological solutions, the worship of mammon that has blocked our ears to the warnings of impending disaster. At the same time we recognize that Christ's community is 'not of this age'. We do not compete for worldly power and territory, and neither should we aim to tear down all corporate structures and institutions in some naive dream of 'anarchy'. New participatory models of democracy and accountable forms of economics are urgently needed. We seek to make space for their emergence knowing that we will always need institutions, organizations and governments to facilitate collective action, and there is no perfect utopia this side of the coming age of true ecological justice and peace. Perhaps it is debatable whether all Powers can be converted, perhaps some just need to be exorcized. But I want to hold onto a sense that the Powers, just like human beings, can be 'redeemed', can become servants of the good, rather than the slaves of destruction. The Powers are always with us and they need to be continually brought under submission to serve the ways of justice and peace, to living in balance with the wonderful gift of ecological life that makes up our home on earth.

Engaging the Powers calls for prophetic action and prayer. Non-violent civil disobedience requires spiritual discernment and discipline. I pray for the church to awaken to the need for more active resistance to the current systems that bring death; to live into our baptism vows to "renounce the deceit and corruption of evil"; to understand that our witness to Christ is fatally domesticated when we restrict it to the private realm, that it is of necessity public and that this will involve sacrifice. What Raj Patel writes about the Gandhian philosophy of non-violent resistance equally pertains to the church:

> One shouldn't underestimate how hard this [engaging market society] will be... to begin with it means regaining an appetite for conflict... Every philosophy of social change has had an understanding of enmity. Gandhian philosophy isn't, as some have reconstructed it, a big tent of beads and incense. Although it's nonviolent, it involves opposition and conflict – tender opposition no doubt, but opposition nonetheless.

Jesus was crucified because he stood in opposition to the Powers. His resurrection and ascension signalled their ultimate defeat, but 'the heavens' in this age are contested and to follow Christ is to take up our cross, to 'regain our appetite' for the conflict and opposition he faced. The post-Christendom church needs to find anew where we are being called to take our stand in Christ's spirit of non-violent love, anger and truth – to name, unmask and engage the Powers of this present age. The community that bears Christ's name is to be a prophetic sign, a site for repentance, resistance, recovery and reconciliation, "so that the Sovereignties and Powers should learn only now, through the Church, how comprehensive God's wisdom really is." (Eph. 3:10 JB)

Vanessa Elston is a pioneer curate with the Church of England in London. She is currently gathering a group to explore what the Spirit is awakening in the heart of the crisis we are in, a journey of repentance, resistance, recovery and reconciliation.

10

Apocalypse now – Revelation for a climate crisis

REV JOHN SWALES

Climate change is real. It is happening now, and it threatens a disaster of apocalyptic proportions. It's as if the horsemen of climate catastrophe are at the gates, and Extinction Rebellion protesters are heralds, warning through civil disobedience of the impending doom.

In light of this uncertainty, we can turn to an ancient book that offers the contemporary church a clear call to have hope and courage in times of crisis: the book of Revelation.

Somebody else's mail

Revelation was likely written at the end of the first century by John, an elderly church leader who had been exiled to the island of Patmos. There he was overcome by the Spirit, and he wrote down his visions. We know the intended audience for the book because John addresses it "to the seven churches in Asia," and it contains individual letters to them. We are misreading Revelation if we think it speaks directly about issues of the 21st century – including the climate crisis. It is not written to us. We are reading someone else's mail.

Throughout the book, there are places where first-century insider information is needed to make sense of the text. Revelation 12 talks about a woman riding a scarlet beast with seven heads, and that these seven heads are the seven hills. For a reader in the first century, it was obvious: the beast and the woman are to be identified with the Roman Empire.

Revelation speaks against the oppressive and idolatrous nature of the Roman Emperor, the Empire and the economic system which supports them. It uses metaphors, describing them as beasts and dragons because they devour the weak, attack the church and bring conflict and catastrophe to God's good creation. Or it describes Rome as seducing the world with prestige and power, but bringing only chaos and destruction.

These images are not meant to forecast events yet to happen. Like much of the prophecy in the Bible, it is more concerned with the issues of the day than foretelling a distant future. The apocalyptic visions relate to Rome, illustrating that the Empire will reap what it sows. However great its power, as it sows violence and exploitation, it will in time reap violence and collapse.

Outposts in the empire

Rome claimed to bring peace and stability, but was known for its brutality. It promised wealth and glory, yet a third of the Roman population were slaves. Its propaganda proclaimed the emperor was the son of God, and that Rome's power was eternal. The Roman poet Statius wrote about Domitian, the emperor who likely ruled when Revelation was written, that he was 'the world's sure salvation' and its 'blessed protector and saviour.'

Rome represents the status quo, and those early churches were called to be outposts of God's kingdom in a world dominated by the Roman Empire. John urges the churches to reject the worldview of Rome and to pledge allegiance to Jesus. Some of them are doing this, clinging to Jesus despite persecution. Others are more compromised, and are told to repent and change their ways. They have colluded with the powers and been unfaithful. "You have a reputation of being alive," John writes to Sardis, "but you are dead. Wake up!"

Although Revelation is not God's word to us, it is God's word for us. It urges us to awaken to the beastly forces that are at work in our world. Where we have been seduced by anti-kingdom and anti-creational forces, powers and perspectives, it calls us to repent. We are encouraged afresh to pledge our allegiance to Jesus.

Pledge your allegiance

Next to the marketplace in Ephesus was an imperial temple for the worship of the Roman Emperor. In order to buy and sell in the market, you had to make an offering and pledge allegiance to the Emperor, and receive an ink mark in return. It is likely that this is what John refers to in Revelation 13 and 14 as the 'mark of the beast'. We can see the challenge for the early church. Allegiance to Jesus is not simply giving your heart to Jesus. It has an impact on day to day life. Revelation calls the church to identify the beastly influences which affect everyday life, choices and economic activity.

For us in the 21st century, it provides a model for how to be a prophetic community:

- We must identify aspects of the prevailing culture which are fueled by greed and exploitation.
- We must repent from our collusion with these forces in our daily lives, and renew faithfulness.
- We must take up our prophetic role to speak truth to power.

The Empire sought to define reality, and Revelation challenges it with a powerful vision of heaven. When the curtain rises, it is Jesus who takes centre stage. It is with Jesus that reality is defined and the true story of the cosmos is known.

To the persecuted or compromised church, Revelation says: look to Jesus. To those in awe of Empire, despairing that anything could be different, Revelation says: look to Jesus.

When we are dazzled by consumerism and put our trust in the status quo; when we are stunned into inaction by talk of climate catastrophe – feeling inadequate or afraid – we too need a vision of Jesus.

For those living in the shadow of the Empire, Revelation offers a vision of the throne of God, as a reminder that the true centre of the cosmos and the true object of worship is God Almighty. All of God's people and all of the creation should pay homage to the Holy One who created all things.

So who sits on the throne? Who is this Jesus to whom we owe allegiance, and who defies the might of Empire?

The eagle, the lion and the lamb

The symbol of the Roman Empire was the eagle, a bird known for its violence, speed and ability to prey on the weak. It reflected the brutality of the Empire, which used its legions to sustain the ruling class in a lifestyle of luxury. Rome was built on fear and bloodshed, conquest and exploitation.

In many ways we still live in a world sustained by violence, and the threat of violence is still used for economic or political advantage. This is the way of Empire, the way of the Eagle. It colludes with Rome, accepting its narrative. For the Christian today, collusion would mean a private faith in a personal Jesus, leaving politics, power, violence and economics unchallenged. Or a church that is a cheerleader for the state, nationalism and empire building.

Another response to Empire is the way of the Lion, which looks for a revolutionary hero who would pounce upon Rome, wage bloody war with the Eagle and become the true King. Drawing on prophesies of the 'Lion of Judah', some early Christians hoped Jesus was this Lion who would lead a violent uprising against the Romans.

Here we see two ways to live with the Empire – collusion or violent opposition. But John's vision of the Kingdom of God subverts our expectations spectacularly. Before the throne of God, John hears the declaration: "see, the Lion of Judah".

Then in an ironic, even comedic twist, John looks to the throne and sees a butchered lamb.

It's not a cruel joke. The lamb is the lion. The lion, with all the authority of kingship, is the lamb. The lamb has authority to speak on behalf of God and bring about God's will. He is worthy of all worship. And this king rules not by violence or coercion but by self-giving sacrificial love.

This is the way of the cross, the way of Jesus, the way of peace.

"Blessed are the Peacemakers," said Jesus. "Love your enemies, do good to those who hate you, bless those who curse you, pray for those who mistreat you" (Luke 7.27). Jesus practises what he preaches. In Jesus we see a God who would rather die for his enemies than kill them. Full of self-giving love, he would rather bless his enemies than curse them; would rather bear the sin of his people than punish them.

God overcomes the forces of death and darkness not through a show of force but through the death, resurrection and ascension of Jesus the Lamb.

The Lamb on the throne is also a model for the church. We should live as patient faithful witnesses to the Lamb, forgiving and loving our enemies. We should be those who speak truth to power – to those who condone violence, the industries which profit from it and the politicians who remain silent. We are called to be those who embody the way of self-sacrificial radical love.

The Unholy Trinity

We too live in a world which tells stories. Consumerism, unrestrained capitalism and individualism promise luxury and privilege, while hiding exploitation and environmental abuse. This Unholy Trinity has taken root in our lifestyles and expectations, making us deaf to the warning of the climate scientists and activists. We may well reap what we have sown.

Patricia Espinosa, Executive Secretary of the United Nations Framework Convention on Climate Change (UNFCCC), said:

> We are witnessing the severe impacts of climate change throughout the world. Every credible scientific source is telling us that these impacts will only get worse if we do not address climate change and it also tells us that our window of time for addressing it is closing very soon.

If things do not change rapidly, the next generation will inherit a world with significantly increased levels of famine, poverty, violence and war. It is difficult to see, without urgent action, anything but the collapse of the current world order.

The Book of Revelation invites us into a different story, calling us to reject the embrace of Empire and the Unholy Trinity, and calling us to pledge allegiance to the loving, liberating and life-giving lamb.

Using metaphor and apocalyptic language, the central chapters of Revelation (6–19) predict that the Empire will enter a season of decline and destruction. During these chaotic times the church needs to act as faithful witnesses.

The book of Revelation does not predict climate change or offer coded messages about current affairs. However, as we live in a time of crisis, the call of Revelation is that we too act as faithful witnesses, followers in the way of self-sacrificial love.

A vision of renewal

In Revelation 21 we see a sweeping vision of a renewed heaven and earth, home to the city of God, with God dwelling among the people. Heaven and earth are intertwined. This renewed earth is physical, and full of non-human worshippers: elephants and eagles, monkeys and macaws. We see community and culture, people from every tribe and tongue. All instruments of war have been beaten into ploughshares and the glory of the Lord covers the earth.

This view of the future should inspire us in the present. Yes, Empire may exploit the earth, and the horsemen of the climate apocalypse may be at the door, but creation matters, physicality matters, and stewardship of earth's resources and biodiversity matter as they are all part of God's future.

It can also inspire us for the future. Life is incredibly painful at times. Death – that intruder in God's good creation, comes knocking – and the unfolding climate chaos will likely see an increase in untimely deaths through shortages, migration and conflict. Yet death will not have the last word.

Isaiah spoke of a time when death would be swallowed up forever, (Isaiah 25.7–8) and in Jesus' death and resurrection we see that the grave does not win. We see a glimpse of the new creation which awaits. Revelation 21 says that God "will wipe away every tear from their eyes, and death shall be no more, neither shall there be mourning, nor crying, nor pain anymore, for the former things have passed away." (Rev 21.4 ESV)

A glorious reunion awaits. Healing and comfort are on the way. The Kingdom is coming. As followers of the Lamb, knowing the great renewal of all things which he will bring about, we are to act as signposts to this kingdom.

We are to live, work, worship and reach out in ways which anticipate what will be. The church is called, however desperate the times around us, to work against suffering, to live as peacemakers, to comfort those who mourn, and to move towards a hurting world full of kingdom hope, joy and peace.

Rev'd Jon Swales is a Team Vicar at St George's Leeds and Tutor in Biblical Theology at St Hild Theological Centre.

CLIMATE CHANGE

MOST RESPONSIBLE, LEAST VULNERABLE

LEAST RESPONSIBLE, MOST VULNERABLE

11

This monstrous shadow – race, climate and justice

AN INTERVIEW WITH ANTHONY G. REDDIE

Climate change is often categorized as an environmental issue. Is it also a matter of global justice?

Some in the global north, particularly in major industrialized countries that have developed neoliberal policies, suggest that the climate crisis is an exaggerated threat, and one they hope to bluff their way through. They know that if they lose that game of chicken with the environment, they won't be the worst affected. They have the resources to make the best of a bad situation.

When climate change impacts upon people's lives, the greatest impact is on those in the global south, and this is fundamentally a justice issue. Failure to respond adequately to the climate crisis is to imperil the lives of significant numbers of people who are already the most vulnerable. They are vulnerable in terms of their poverty, the inequalities of global trade, and other aspects of globalization that have already affected them. In addition to all of those problems that they are already battling with, places directly impacted by climate breakdown are now going to have to battle for their very existence.

Given the differences of the global North and South, is there a racial dimension to climate justice?

Absolutely. We don't see it because the media doesn't tell us those stories, but let me give you an example. My day job is with the Council of World Mission, and a third of our member churches are in the Pacific. Some of those Pacific islands have already been evacuated, because they are low lying and sea level rise has made them unliveable. Think about how we would feel if our ancestral lands, the places where we have invested our emotional and physical energy over centuries, suddenly disappeared and we had to evacuate. Imagine how catastrophic that would be for our identity and self-understanding. And yet somehow it happens in other places far away and it goes unreported.

I recently heard a presentation from the climatologist Dr Leon Sealey-Huggins, on the apocalyptic nature of climate justice. He focused on the Caribbean – where he and I both come from. There was something he said that chilled me to the bone: that while some powerful elites in the West are climate deniers, others accept that it will happen, but are working on the basis that some collateral damage is acceptable.

There are powerful people who can countenance people dying and cultures disappearing, if that's what it takes for everyone else to continue – if only for a period – with some sense of normality. We still want to grow, to consume, to pursue rapacious capitalism. If some Caribbean or Pacific islands disappear, that's just the price we pay for 'more civilized' people to continue their lifestyles unchanged and unchecked.

What was so chilling to me was the thought that some of the people who think this way could also be right wing evangelical Christians. The notion of sacrifice, especially if it involves someone else, is something that they can build into their understanding of the world because of the way we theologize around Jesus' sacrifice. Jesus says 'if you want to follow me, pick up your cross'. And of course, we're being very selective about who should pick up the cross, and who should be willing to sacrifice. Inevitably it's not the people who are using the rhetoric – it's imposed on others who don't have the choice. And so there's a huge race element involved that we are not seeing.

Do we push away these elements of climate justice because they make us uncomfortable?
We don't like to talk about race full stop. It's often the elephant in the room. We try to abstract it and deflect it by looking at other things, but that means we miss underlying causes.

The climate catastrophe has been triggered by the rise of capitalism, exploiting resources and using them remorselessly for growth and for profit. That process isn't new. At the moment we're doing that to the environment, but we used to do it to people. As Sealey-Huggins also points out, the slave trade was underpinned by rampant greed and profit, the displacing and exploiting of bodies. Once we'd done that, we moved on to the environment, grabbing fossil fuels to drive growth. The impact of climate change on black and brown-skinned people comes on the back of 500 years of exploitation of their bodies anyway. It's a compound disaster, adding one injustice to another.

This is so deeply etched into our way of being that it's this monstrous shadow that we don't even talk about. Because if we do, we have to go right back to the very basis of why some countries are rich. We talk about the industrial revolution, and how it transformed Britain from an agrarian society to an industrial powerhouse. We never talk about the fact that the capital for industrialization came from slavery, from the exploitation of the people who are now most imperilled by climate change. The people who suffered 500 years ago are the same ones suffering now.

As a theologian, how can our theology either help or hinder our understanding of these issues?
We have to dispense with one theology and pick up another. We need to dispense with what is called 'dominion theology'. That's the idea that God has given us dominion of the earth for us to do what we like with it. It's a misreading, but even if it were accurate, it needs to come to an end. We are custodians of something that we did not create. Once we exploit it we can't recreate it.

On the positive side, there is a helpful perspective in what is called 'process theology'. Process theology makes a distinctive switch in thinking about God and creation. Historically we've thought of God as being separate from creation, as this being who is not defined by space and time.

Process theology asks whether God is in fact not external, but is locked into this with us. And therefore, when we are mistreating the creation we are mistreating God. Now obviously this can slip into pantheism or other problematic areas, but I find it valuable because it gives us a powerful context for this unique creation, this special place in the whole of the cosmos. We can think of God with us, here in the mess as an active participant helping us fix the situation. This might not be considered orthodox theology, but I can live with the 'heresy' if our 'orthodoxy' is leading us to kill creation!

Alongside new theological understanding, what can the church do practically to address these inequalities?
If we can't change the whole world, what's the little we can do that's within our control? How we help congregations to think about their buildings, and how they can operate in more ecological ways? This might seem small, but if you multiply those actions across all the churches, that in itself would be significant. People can often feel despair over climate change, and knowing we can make a difference, even at the local level, is empowering.

On a bigger scale, I'm reminded of a colleague at Coventry University who says that churches are good at social welfare, but not social action. We're good at helping out when there's a problem, but not so good at using our influence to lobby politically on systemic injustices.

Bill Clinton famously said 'it's the economy stupid', and as Christians we need to say we will support politicians who say 'it's the environment, stupid'. We believe in a God who has given us this creation, a gift bestowed upon us for safekeeping for future generations. What would happen if we, as a movement of churches, supported candidates or parties that put the environment first? More economic growth is irresponsible. Consumption is irresponsible, and we need to change. We should support visionary politics that tells the truth about these things.

That said, history tells us power is not conceded. You have to make change happen, and direct action – peacefully being the best way – can make a point powerfully enough to change the discourse. We cannot continue with business as usual, as if nothing is happening. People are dying and cultures are disappearing as we speak.

A hundred years ago Emmeline Pankhurst and the Suffragettes were using tactics of disruption to make a political point. In some regards race was involved there too, because the movement was predominantly entitled middle class women. That did not invalidate what they were doing. There are similar questions around Extinction Rebellion, but that does not invalidate what they do either. It just means that we have to have a broader coalition and ways of operating.

On that point, what advice would you offer to a movement that wants to diversify and be more inclusive?
I'll admit, I would find it hard to be part of Extinction Rebellion. For me and for many black people – in getting yourself arrested, can you expect a certain level of safety, or trust in a democratic process? There have been too many black deaths in police custody, for which no police officer has ever gone to prison. I am not disposed to make myself vulnerable to arrest voluntarily. That does not invalidate those who are prepared to do it.

What we need is a range of strategies, so that people can participate without assuming that it's one approach for everyone. I would find it difficult to get myself arrested, but there are other ways that I can contribute to this movement.

It starts with dialogue. Extinction Rebellion can begin conversations with different black communities and work out what our entry point could be. We should ask how we can support each other in the different tactics that we use, acknowledging that we are all on the same side, because it's the one creation that all of us share.

It's not about saying 'let's dialogue with black communities so that they can come and join us', because that presumes that power lies with you. We mustn't assume that others aren't doing anything either. A better question is to enquire about what they are doing, and how can we support that? Race issues are complicated, but there's a bigger issue that we need to come together around, because ultimately this is about capitalism and growth, and a status quo that has to be critiqued and challenged and changed.

Anthony G Reddie is an activist theologian and educator, author of 17 books, and the founding editor of the Black Theology *journal.*

12

The strategy of a non-violent uprising: disruption and sacrifice

MARTIN NEWELL AND
HOLLY-ANNA PETERSEN

At Christian Climate Action, we are often asked why we break the law. Others ask "Why do you insist in getting in the way and blocking roads? Don't these things just annoy people and make them dislike you?" These are fair questions, and the Extinction Rebellion strategy is controversial. The answer is that in our actions we try to use principles of change which have been shown to be effective in previous non-violent uprisings – principles such as *sacrifice, disruption* and *non-violence*.

Perhaps the principle that people presently appear to understand the most is that of non-violence. It is the one most readily seen as a moral principle, though this has not always been the case. When we as a nation have been faced with other national emergencies, such as risk of attack or invasion, the anger that is generated has quickly moved to advocacy of violence. In these situations, many Christians and church leaders have been silent or even supported use of violence. Now we are faced with an emergency of even greater force. It is not just one country or region that faces invasion and the destruction of their 'way of life'. It is the whole world. And not just human life, but all the life of God's creation is being threatened. We are being 'invaded' by carbon and other greenhouse gas emissions, through power structures that cause and profit from the destruction. And some have far more power to maintain or to change the status quo than others do.

Of course, few of us in the world's richest countries are entirely innocent. At CCA we recognize and confess our own complicity, and work for change in a context of global justice, seeking to repair past and present wrongs as part of the healing process. If we do not act in solidarity with those at the margins, responses to the climate emergency may exclude them again. When the world fully wakes up to the enormity of the climate emergency, some may be tempted to call for repressive solutions. It is important to understand why violence is not the answer – and why active non-violence is.

History is full of incidents where those seeking change turned to violence to get their way. Yet research shows that active non-violent uprisings are more effective than violent ones. In *Why Civil Resistance Works*, Maria Stephan and Erica Chenoweth show that over the last century, non-violent resistance movements have been about twice as effective as violent resistance movements. We believe that this power of non-violence was taught and practiced by Jesus, as described in other chapters in this book. The non-violent resistance strategy of Jesus has been analysed by Ched Myers, Albert Nolan, Dorothy Day, Walter Wink and many others. However, non-violence is not easy. The post-WWI pacifist movement saw the non-violence Gandhi taught and practiced as "the moral equivalent of war", requiring a similar level of suffering, discipline and self-sacrifice for the common good of all.

Still, non-violence is one thing, disruption another. The question remains: "Why block the roads or break the law?" The answer is that it is a strategic decision, based on the research by the movement's founders. Disruption and sacrifice are fundamental elements of this strategy.

First, disruption is important because it gets noticed. The general public already have many other things to worry about, whether serious or trivial. Government and business are likewise focusing their attention elsewhere. Sadly, protestors standing by the side of the road or marching around for a day and going home does not break through – even if participants are numbered in millions, as they were for the march against the Iraq war in 2003. To get people's attention, something has to have an impact on them and affect them emotionally. The second thing that disruption does is directly push those in power to make changes, through the economic costs imposed. When the cost of the disruption exceeds the cost of doing what is being demanded, those in power often give in. That is why trade unions and workers use the tactic of going on strike.

To cause disruption, XR has been carrying out actions such as blocking roads, and even hoping to clog up and disrupt the criminal justice system with large numbers of arrests and court cases. A common criticism is that this disruption will build opposition, and there is clearly some truth in this concern. However, the research by Stephan and Chenoweth has shown that only 3.5 per cent of the population need to actively support an uprising for it to be successful. This means that XR does not need everyone on board for the movement to achieve its aims.

Another criticism is that disruption targets the wrong people and places, and certainly we will never get everything right. We should of course be aware of who is disrupted, and how often and for how long. Yet individual actions should be assessed on their own merits, and a misjudged action does not mean that the principle of disruption is ineffective. At CCA we plan actions prayerfully, debrief afterwards and do our best to make sure we learn from any mistakes. The gospel calls us to take the risk of acting, rather than the comfortable safety of apparently avoiding mistakes by doing nothing.

If disruption still seems like a difficult principle to stomach, it is worth remembering that Jesus disrupted the lives of ordinary people, as well as the economic system, when he turned over the tables of the bankers and the traders in the Temple. In addition, he disrupted their – and his – place of worship. Many of Jesus' contemporaries would have regarded the activities of the Temple worship and system as the most important activity of their time and place.

Yet disruption alone will not bring the change that is needed. We also need to change hearts and minds. One thing that changes hearts and minds is sacrifice – the willingness to suffer voluntarily for what is right. This is a basic Christian theme – the way of the cross, ultimately even of martyrdom. It was the willingness to voluntarily suffer for love and righteousness that brought redemption into human history and changed the world. Fully half of the

Gospel of Mark is the Passion narrative of Jesus' suffering and crucifixion. It was this, of all the events of his natural life, that most astounded those who were with him. Even someone rising from the dead was, on its own, insufficient to start a revolution. (Jesus is not the first person in the Bible to come back to life.)

Willingness to suffer for a cause certainly convinces people of the sincerity of the one who is willing to suffer. When the suffering is seen to be unjust, it also builds sympathy. There is a global justice angle here too. When we suffer for the cause of climate change, we acknowledge and stand alongside, and in solidarity with, those at the sharp end of climate breakdown – the displaced, the indigenous communities, the most vulnerable.

Non-violence, disruption and sacrifice are by no means an exclusive list of principles of change, but they have been core to successful social justice movements of the past. Their effectiveness is why they have been used over the last century. Of course, the climate emergency is different to previous non-violent revolutions, in its circumstances and its aims. When privileged people take action, it does not have the prophetic immediacy of the oppressed taking back power for themselves. That should make us wary of drawing direct comparisons, though it does not invalidate XR's actions. Normal campaigning has not brought about change at the scale or speed required. We need a non-violent mass uprising because it is the only thing that might work, given the time available to us and the urgency of the threat.

Of course, another distinctive of our activism as Christians should be humility. We cannot claim to have all the answers, or any guarantee of success. We must be open to being challenged, to learning and improving. Yet we live at a time of global emergency. It is time to act.

During the apartheid era in South Africa, the churches for a long time sat on the sidelines debating the respective merits of violent and non-violent resistance. Walter Wink challenged them. He told them that they weren't going to use violence, so they had better stop debating and support the active non-violent resistance.

Our context is different, but perhaps the appropriate challenge today is "come on, enough talk. Time is short. Wake up to the emergency and get out on the streets". Because much as it is clear that non-violent movements are more effective than violent ones, they are also more effective than passivity and the usual rounds of campaigning. What is needed is a mass non-violent uprising, the moral equivalent of war, to urgently transform the system causing this climate breakdown. In this emergency, we do not have the time to wait.

Holly-Anna Petersen is a mental health practitioner and has been a member of Christian Climate Action since its inception. Martin Newell is a Passionist priest and founder of Christian Climate Action.

13

How indigenous perspectives offer hope to a beseiged planet

RANDY WOODLEY

Change your lenses, please. Okay, maybe you can't simply change lenses right now, but would you at least notice the lenses you are currently wearing? If you are like, say, 99.9 per cent of us in the US, you have been influenced by a very particular set of perspectives that interpret life from an Enlightenment-bound Western world view.

All of our lenses have various perspectival tints, but Western worldviews seem to have several in common, including the foundational influence of Platonic dualism, inherited from the Greeks. This particular influence absolutizes the realm of the abstract (spirit, soul, mind) and reduces the importance of the concrete realm (earth, body, material), disengaging them from one another. In dualistic thinking, we are no longer an existing whole.

Western world views tend to have some other related assumptions – such as hierarchy, extrinsic categorization, individualism, patriarchy, utopianism, racism, triumphalism, religious intolerance, greed and anthropocentrism. However, the influence of dualism empowers these other concerns.

What difference would it make if life were viewed instead as a fundamental whole, if the earth itself were seen as spiritual? How would such a world view square with Jesus' approach to such matters?

A few encouraging facts as we approach these questions:

- Most of the rest of the world does not understand life through a Western worldview. We in the West are the anomaly.
- Jesus was not an Enlightenment-bound Western thinker. He thought more like today's premodern Indigenous people.
- Not one writer of the scriptures saw life through a Western lens.
- Indigenous Peoples of the world have an advantage over Western thinkers in that there is still enough premodern worldview intact among North American and other Indigenous people to relate to the premodern Jesus and the premodern scriptures. They can bring new kinds of hope to today's earth climate crisis, if we allow it.

Jesus understood humanity's relationship with the earth differently than we do. He spoke to the wind, to the water, and to trees; closely observed the habits of birds, flowers, and animals; and called his disciples to model their lives after what they saw in nature. In Matthew 5, during his Sermon on the Mount, Jesus said, "Do not say, 'By heaven!' because heaven is God's throne. And do not say, 'By the earth!' because the earth is his footstool" (34–35 NLT).

Jesus was making a point about making vows, but one of the many by-products we see from this short exchange (and from his whole life) is Jesus' view of the whole world, including earth and heaven, as sacred. Jesus understood the balance between the earthly and the heavenly realms, and he certainly understood the relatedness of both ("on earth as it is in heaven"). Jesus was firmly planted in the construct that "the earth is the Lord's and all that is in it."

The predominant themes and subject matter of Jesus' stories were natural, such as fish, flowers, birds, sheep, oxen, foxes, earth, trees, seeds, harvests and water. There were many mechanical inventions during Jesus' time, but the record reflects he paid little attention to them. His was a world of keen observation, where God was wondrously alive and at work in creation.

In Jesus' world view, he laid to waste the fallacies of Platonic dualism that exist in our modern era and that presume the earth or the body or anything earthly is less spiritual than the mind, the spirit, or things more abstract. To Jesus, as it should be to us, the earth is wholly spiritual, as are our bodies.

Earth out of balance

If we remove the influence of Platonic dualism from our world views, we find it difficult to view human beings as being over all other parts of creation. Instead of a relationship where nature is below us, we should be stewarding with, or co-sustaining, all creation.

Each area of creation is working with the Creator to maintain earth's balance. The rain and snow, oceans and sun all sustain life on earth. Animals regulate each other within their various natural cycles. Plants provide oxygen, food, and shelter for all of creation to coexist together. As human beings, we are co-sustainers with the rest of creation to ensure the abundant life for all creation the Creator intended. And God said that is good.

When considering our relationship to the earth, Christians will recognize that Jesus is earth's creator and sustainer (John 1, Colossians 1, Hebrews 1), and we simply cooperate with him in these tasks. Our job, as humans and as Christians, is to maintain the natural balance God set forth through Christ.

Unfortunately, things are out of balance. By allowing past and current exploitation of the earth and earth's resources, we are now reaping the consequences via the current climate catastrophes. Earth's topsoil is disappearing; coral reefs are dying; glaciers are melting; aquifers are not being recharged; animal, bird, fish and plant species are dying at exponential rates. We are experiencing a steadily increasing amount of severe weather patterns through hurricanes, flooding, tornadoes, forest fires, landslides, droughts, and snowstorms – and the costs of these disasters continue to rise progressively each year, into the hundreds of billions of dollars. Earth is out of balance, and as a result all God's creation is in peril.

The road to restoration

Christians have evaded the responsibility of earth-care, due largely to the adoption and influence of a worldly philosophy. By ignoring the earth's problems, are we not responsible for dismissing the things important to Jesus in favour of our own selfish interests?

Indigenous peoples have historically been condemned because we view our relationship with the earth to be very sacred. Like Jesus, indigenous peoples understand their relationship with creation as paramount to the abundant life God intends for all humanity. In other words, to be human is to care for creation. If we want to live our lives together in abundance and harmony, and if we want future generations to live their lives together this way, we must realize we are all on a journey together with Christ to heal our world. Earth healing will take cooperation from all of us to solve the problems.

The single conceptual integration of land, history, religion and culture may be difficult for many Western minds to embrace. For indigenous peoples, this integration is often explained as a visceral 'knowing', or as somehow embedded in our DNA.

This feeling we have of ourselves as a people, including our history and cultures, being connected to the land is perhaps the single most glaring difference between an indigenous native north American world view and a Euro-Western worldview. Yet if we are all to survive the 21st century, things must change so that our Euro-Western friends can sense a similar connection. How does such a paradigm shift happen?

A world view both indigenous and biblical

Christians cannot merely leave such matters in the hands of well-meaning secular environmentalists. Although everyone should be deeply indebted to those on the frontlines of the environmental movement, many current initiatives are only helpful in the short run, because they focus on preserving earth and water for a particular use.

Unless there are guiding values that become rooted in a familial love of creation, these short-term initiatives simply may represent a more sanitized version of utilitarianism: using the earth without deeply loving the earth as sacred creation. Utilitarianism (using the earth out of self-interest, for good or bad) has been, in part, the problem.

This is where indigenous people can be helpful. Many indigenous peoples understand:

- *Creation exists* because of a Creator.
- *Life is intrinsically valuable because* it is a gift from the Creator and, therefore, it is sacred, meaning that sacred purpose is crucial to our existence.
- *The role of human beings is unique*, and humans relate to the rest of creation uniquely. This includes restoring harmony through gratitude, reciprocity, and ceremony between the Creator, humans and all other parts of creation.
- *Creation does not exist to be ignored* in isolation, but creation is the Creator's first discourse in which humanity has a seat of learning and in which the discourse is continuous.
- *Harmony is not simply understood as a philosophical idea*; it is about how life operates and the only way that abundant life can continue, if life is to be lived as the Creator intends.

We are now at a point in human history when we must realize that the industrial age has written a check to our world that has insufficient funds. Only a world view that encompasses the interconnectedness between Creator, human beings and the rest of creation as one family will sustain abundant life.

Such a world view is fundamentally both indigenous and biblical. If we are wise, we will protect Christ's creation, for this creation is central to God's investment in us.

The rights of nature

There are a number of obstacles that the Western world and worldview will need to clear in order to rebalance and preserve creation, but there is one great and relatively timely action that could move us quickly down the road to restoration: We can pass laws to protect Christ's creation – not just because of the current climate crisis but because we love what God loves and we want to understand the world more like Jesus, earth's creator and sustainer.

Following the lead of indigenous people's movements around the world, we can enact laws and constitutional amendments to protect the earth, as did Ecuador and Bolivia. More than three dozen US municipalities have adopted similar 'rights of nature' laws and regulations, including Pittsburgh, the largest US city to do so.

Bolivia's 'Rights of Mother Earth,' as it is sometimes called, grew out of the World People's Conference on Climate Change held in Cochabamba, Bolivia, in April 2010. At that conference, more than 35,000 people from 140 nations adopted the Universal Declaration of the Rights of Mother Earth. (You can find more on the rights of nature movement at <The RightsofNature.org>.)

Bolivia is one of many countries struggling to deal with the climate crisis and its weather anomalies, such as rising temperatures, melting glaciers, numerous floods, droughts and mudslides. Bolivia, like many other nations, is a battleground country between the rights of Indigenous peoples, especially landless peoples, and a corrupt corporate state, similar to the situation in the US, where corporations are legally protected 'persons' but the earth that supports them has no real voice or rights.

To remind us of our intimate connection with creation and to break dualistic thinking, 'action' must become an important word in Western people's vocabulary. Euro-Western people might also consider developing new ways of expressing their thanks through outdoor, earth-honoring ceremony. Through expressing gratitude in ceremony, Indigenous Peoples reveal to others and themselves the connection between Creator, human beings, the earth and all the rest of creation.

A foundation of Native American ceremony is gratitude for the relationships that exist. Euro-Western people can rediscover what their indigenous ancestors once knew and, in so many ways, reclaim some of their indigeneity once again. To move ahead – perhaps simply to survive – we must all be connected to Christ's creation in harmony.

Throughout the gospels Jesus gave creation a voice. We should do the same. Would you ask your spouse to remain silent or ask them to not make their needs known?

We are in a relationship with the earth and all of earth's creatures. We must make the earth's voice known and protect it by recognizing the earth's rights. Unfortunately, we've waited until this late hour to realize the sacredness of this relationship. Let's not delay until it is too late. Jesus is waiting.

Dr Randy Woodley is a pastor, professor and podcaster of Cherokee descent. He is the author of Shalom and the Community of Creation: An Indigenous Vision.

Reprinted with permission from *Sojourners* (800) 714–7474, <www. sojo.net>.

14

Is it too late?

STEFAN SKRIMSHIRE

In a book entitled *Time to Act*, should we fear, or flat out refuse, this further question: is it already too late to take action on the climate and ecological emergency? Might the very question justify doing nothing in the face of many urgent choices that lie ahead of us? Alternatively, if we take the question seriously, then how ought the Christian activist respond?

This is not a new question for theology. One of the pioneers of eco-theology, John Cobb Jr., published *Is It Too Late?* in 1972, the same year that the Club of Rome published *The Limits to Growth*. Both were sounding the alarm about agricultural degradation, population growth, pollution and rising inequality. Yet whereas Cobb's warning probably seemed radical then, today it seems almost mainstream. Today it is quite common to hear that the fight against climate change has been lost. As a researcher and participant in the climate action movements of the past ten years, this doesn't surprise me. The language of those movements has always focused on deadlines for taking action, or thresholds that we can't afford to breach. We've been told that we have 'ten years left' for longer than a decade now. The 'safe' threshold of atmospheric carbon dioxide was given at 350 parts per million, and we reached past 400 in 2016. In 2019 scientists told us that key tipping elements in the Earth system – which may lead to a cascade of others – have already been breached.

Can the energy of our activism survive this culture of 'deadline-ism', as Mike Hulme has called it? The past two years of international climate protests would suggest that rebels are not only undeterred by such language, but even emboldened by it. Yet still, there are good reasons for suspicion. First, it is never clear to what the 'too late' refers. Words like 'runaway', 'dangerous', 'catastrophic' and even 'irreversible' are notoriously subjective. For some, it was too late a long time ago; for others, we have everything yet to lose.

Second, we should really be asking, too late for whom? So often the language of climate emergency speaks about 'our' last chance for 'survival', speaking about the human species in an abstract sense. As if millions of human and more-than-human others were not already subject to catastrophic climate change or the violence of extinction. Hasn't the world already ended, several times over, in the history of industrial ecocide, rising sea-levels, and settler colonialism? Christian activists know – or ought to know – this very well, as charities like CAFOD, Christian Aid and A Rocha bring to our attention the climate catastrophe that is continually befalling the poorest of the Earth. Nevertheless, the tendency of our activist language can be to think in terms of a temporal binary: it is / it is not 'too late' to stop catastrophe from occurring.

Christian activism can easily become swept up in this

language. When it does so, it tends to seek a theology that can heighten the sense of our exceptional moment, the moment 'just before' it is too late. The New Testament is full of expressions like this: "The appointed time has grown short" (1 Cor 7.29 ESV), writes St Paul, meaning that with the life, death and resurrection of Christ, we stand at the threshold between the old world and the new. Jesus himself, while deterring his disciples from calculating the time and place of the end-times, implores them to remain vigilant, "because you do not know on what day your Lord will come" (Matt 24.42).

To people who have concluded that it is too late to halt the climate emergency, it can seem as if time is 'becoming short' in a literal sense. It is difficult to know how to interpret the role of Christian action within this belief. For some, the important question is no longer whether it is too late for human action, but whether it is too late for God's. But even if the answer derived from our faith is an emphatic no, does that mean that we are finished with a theology of activism, of co-operation between God and the Church? There is the danger that a theology of hope devolves into an act of patient waiting for God's act of redemption, the more that our global crises seem to describe apocalyptically the other side of the 'too late' binary vision. This is why the perception of Christian faith by those on the outside often seems so antithetical to climate action. It is tempting to think that it is too late for sinful humans to fix the mess, and all that remains is to hope and pray for God to "make all things new" (Rev 21.5).

The experience of Christians and others of faith who are

rising in huge numbers to meet the climate and ecological emergency, tells a much richer way of thinking about our present moment. The spirit of that commitment seems to be: it may well be too late to halt some very dire future scenarios indeed, and for some concrete situations in the world, there is no future. Yet it is especially in such a context that the Christian is called to take action, and in acting, to resist despair. The activist must loosen their grip on knowing with any certainty whether their actions will 'succeed'.

Talk to activists around the world who are doing all they can to create carbon neutral societies, and you may find that secretly, yes, they believe it may be too late to reasonably achieve that scenario. Yet to strive for that world embodies more than a reasonable hope; the striving itself embodies the world in which they believe. The sense that it might be too late to take action is a legacy of secular, not religious thinking. To worry that it might be too late betrays a desire for certainty. We would like to have reasonable grounds for our actions making some difference to the world. We would like to easily distinguish actions of 'mitigation' (those which prevent a situation like runaway climate change) from those of 'adaptation' (those which simply prepare oneself for that unavoidable situation). Christian activism expresses a different logic, inspired by faith rather than certainty.

The most powerful of the Bible's ecological references are those that chasten the human desire to predict its trajectory, or to know it as God knows it. From the book of Job, to the Gospel passage cited above, the human desire to understand the fate of the world is put in its place. It is not, it seems, the role of the human to decree when things are 'too late'. We may well have knowledge that the near future will be – and already is – catastrophic, indeed a real 'end-times' for certain people, creatures and lifeways. Yet it is only faith that can truly leave open what is possible in such a world, through our actions.

Deep Adaptation

This is the spirit by which we believe Christian activists might engage with Jem Bendell's much discussed manifesto for 'Deep Adaptation' (2018). Bendell concludes from a survey of climate reports that "there will be a near-term collapse in society with serious ramifications for the lives of readers". He suggests that many of us (certainly academics) are engaged in some level of profound denial of that fact. The work of deep adaptation means taking seriously what it means to redirect all of our pursuits and values to meet this reality with as much care and compassion as possible. Bendell's language is controversial for many people. Declaring "there will be" is like declaring "it is too late" – again we would want to ask, how do we know what this 'collapse' will look like? For which people? There are already citizens in parts of the world – for instance Pacific islanders – who face the near certain disappearance of their homelands from sea-level rise. To talk of social collapse as inevitable is to disregard the society-wide work by which such communities respond to external crisis, and have been doing at least since colonial times. The idea of inevitability also disguises the fact that collapse is in in many contexts due to, or exacerbated by, social inequalities and injustices. There is nothing inevitable about injustice. It is never too late to oppose it; the gospels are pretty clear on this.

However, the work of deep adaptation that has inspired countless workshops, talks and discussions, might have an important role, to which Christian activism clearly has something to offer. For adaptation is surely also a form of activism, inasmuch as it is an invitation to transform our ways of living. Asking what things we will need to relinquish, or restore, or be resilient to in the light of social collapse, invite an exercise of imagination which has, arguably, always been at the heart of Christian ethics. Those ethics were developed in a world in which catastrophe loomed large. First in the form of (Roman) imperialism, then at the collapse of that world order. For many it seemed that the world was indeed ending. Loving God by loving one's neighbour; giving away one's attachment to earthly possessions; considering all things held in common – these familiar injunctions for the Christian are especially significant for having emerged from a place of collapse, and radical uncertainty with regard to the future.

So it is with the activist facing the question of whether it is too late to take action. Perhaps to have faith in a time of collapse is to leave open (to God, perhaps) what the purpose of our actions are. Showing solidarity with those who are suffering more than us; halting or slowing down our acts of violence against the Earth; helping ourselves and others to lament for what has already been lost; fighting for a more just transition to a carbon free world. When you look at it from the perspective of faith, the lines between mitigation and adaptation become blurred. This is a good thing. Because to believe that "the time has become short" means not that time is running out, but that the time of action, the "favourable time" (2 Cor 6.2), is always now.

Stefan Skrimshire is an Associate Professor of Theology and Religious Studies at The University of Leeds, researching and engaging with climate activism.

15

How dare they!

CAROLINE BECKETT

How dare those grubby activists
Disrupt our peaceful lives
With scary stories telling us
Our world might not survive!
How dare they block the traffic flow!
How dare they make a fuss!
How dare their battle for our world
Impact the rest of us!

How dare they bang a different drum!
How dare they make a noise!
How dare they make us hear their case
And not give us a choice!
How dare they chalk on monuments!
How dare they clean them up!
How dare they hold a silence!
How dare they not shut up!

How dare they look so different!
How dare they look like me!
They're clean – they must be cheating then!
They're hemp-smelling crusties!

How dare they!

They're irresponsible parents
Who deny their children treats!
They're just a bunch of hypocrites -
Look, that kid's eating sweets!

How dare they make us feel afraid
And dress in scary red!
How dare they stir our feelings up
Fasting or playing dead!
How dare they dance and laugh and sing,
Be witty, funny, arty!
How dare they groove to samba drums
And eat and drink and party!

How dare they not be perfect!
How dare they still eat meat!
How dare they tell us that they're scared
While blocking up the street!
How dare they once have flown somewhere.
What? Drive a car? How dare they!
How dare those who would change the world
All be so ordinary!

They should all be arrested!
They're thugs committing crime!
They're just grandparents, geeks and kids!
They're wasting police time!
They're just a bunch of scroungers!
Looking to lounge and shirk.
Wait, what? That group are doctors!

The Head

They ought to be at work!

They're just having a good time -
A selfish get-together.
Old folks and little children
Shouldn't sleep out in this weather!
They're just a loud minority,
Put your fingers in your ears.
And la la la don't listen
Till the nuisance disappears.

Don't listen to the scientists,
The crusties or the activists,
The Christians, Muslims, Buddhists, Jews,
The Pagans, Humanists, Hindus,
The teens or the celebrities
The hippies or the nobodies,
The children or the grandparents,
The doctors or your common sense...

Or...
You could maybe *not* be sheep
Drifting to crisis half asleep,
Condemning those who dare to sound
A loud alarm to bring you round!

Revd. Caroline Beckett, Vicar of Brightlingsea, mum of two teens who need a future, wife of Mike Beckett, Green Liberal Democrat; grassroots campaigner for social justice and action on climate change.

NOW !

CHRISTIAN CLIMATE REBELLION

CCA members on the launch day of Extinction Rebellion.

THE
HEART

16

Burnout is not the sacrifice we need

HOLLY-ANNA PETERSEN

When we hear stories from social justice movements of the past, it can sound as if activists were equipped with superhuman courage, confidence and determination. These heroic figures would do 'whatever it takes' and 'will not rest' until the injustice is over. This narrative certainly gets our adrenaline pumping, and it ignites a sense of restlessness in our bones. However, I think the reality is actually far more inspiring.

The truth is that rebellions are not run by superhumans. They are made up of ordinary, broken people just like everybody else. Most of us are taking part around our jobs, around our families and around our other life commitments. Rebellions are hard graft. What I find inspiring is the fact that even though we are far from perfect for the role, we have stepped up anyway – because if not me, then who? If not now, then when?

What this gritty reality means is that we need to recognize our fragility and limitations. Black and white statements like we should 'do whatever it takes' may have a romantic pull. They can also give the illusion of bringing some clarity and certainty in a somewhat confusing world, but we need to acknowledge that going forth all guns blazing is not good for us, and neither is it good for our rebellion. As with many things in life, the best way forward lies in balance – between intentional action to bring about the change we need and making sure that we have rest to revitalize.

There is a lot of research outlining the importance of rest. It is well-known that rest plays a vital role in our emotional and physical wellbeing. It has also been shown to enhance our performance – such as in creativity, problem solving and productivity. With this wonder-cure having such far-reaching positive effects, it comes as no surprise that it is a well-grounded biblical principle. From the first pages of Genesis, God weaves rest into the fabric of our world. We are told that even our all-powerful creator set aside time from his creating for reflecting on his work and having a day of pause. This pattern continued when God became flesh. Although Jesus was clearly a man of action, he also spent a lot of time alone, in quiet, reflecting and praying.

Two important aspects of rest are captured in Psalm 46.10: "Be still and know that I am God". These eight words outline elements of both practical rest and emotional rest.

Practical rest

This involves us establishing a regular rhythm of intentional acting and resting which is built into all we do. We need to understand the importance of these rhythms which permeate through all the "dust of the earth" – from each cell in our bodies, to the great seasons of the hemispheres.

Every person in a group needs to be able to take time to

rest. Doing this well involves members striking a balance between two powers – that of saying 'yes' and that of saying 'no'. It can be a scary thing saying yes – especially when it is for a task which is unfamiliar to us. It can be tempting to sit on the sidelines and wait for someone else to opt in, perhaps someone we consider better suited to the task. However, participating in a rebellion is something that is new to most of us, so the majority of tasks involved are going to be outside of our comfort zones. In this climate movement, we are all 'crew', meaning that if we want the vision to come true, we must all participate. Exercising the power of saying yes gives other members of the group the reassurance that they can rest when they require it – allowing the group as a whole to be more sustained.

On the other hand, saying yes when we already feel overloaded can lead to a slump in our productivity or even burnout. It's important that we take time to listen to ourselves and learn where our limits lie. We shouldn't take on another action, project or task if we already feel overwhelmed, and we should let others know what we need a period of rest.

'No' is not a sign of weakness. Consider it a positive choice to rest. Then, the next time we say 'yes', it will be from a place of strength.

Emotional rest

Just as important as practical rest is emotional rest. There is no point scheduling in some down-time if we are using that to dwell on self-defeating thoughts that leave us feeling physically and emotionally exhausted. It's a strange element of our culture that being hard on yourself is considered a sign of strength! It's not – it's a sure way of making us less resilient to life's inevitable knocks and bumps.

There are always going to be things that we wish we could have done differently – be that a poorly executed training event or an ill-planned action. Reflecting on these together through structured debrief sessions is a way of learning together as a movement. Yet it's important to acknowledge that we don't grow by beating ourselves up. To thrive and stay energized in this rebellion, we need to be kind to ourselves. We can do this by forgiving ourselves for the times where we inevitably make mistakes, and taking regular intervals to celebrate the good that we have done – just as God did when laying the foundations of the Earth.

Another thing that can disturb our emotional rest is the number of elements of what we do that are outside of our control – be it the weather, police tactics or the responses of the media and the general public. Uncertainty and unpredictability are the primary things which feed our anxiety levels. At these times we can be tempted to get short-term relief from our tension by trying to stamp out uncertainty. This can manifest in us overpreparing, trying to exert control over others or spending our time worrying – thinking through the endless potential negative scenarios that could happen. While these might feel like solutions, the sad truth is that we are never going to be able to eliminate uncertainty. It is intrinsic to life. Attempts to do so will inevitably leave us feeling exhausted, and less resilient in the moments when we need to be responsive.

Instead of grasping for certainty, we need to learn to accept uncertainty and sit with the unsettling feeling that it brings. One of the beautiful things about being a follower of Christ is that we can hand that uncertainty over to God and draw comfort from the Holy Spirit. Rather than worrying about all the things we cannot control, we can fall back on our calling: as Christians we are called to be faithful, not successful. We need to focus on whether our actions are done prayerfully and with love. We cannot be certain, but we can act with good motives, taking responsibility and being ready to learn, whatever the outcome might be.

Brothers and sisters, we are not superheroes. We are ordinary – but with courage and a regenerative culture that respects our rhythms of action and rest, we are capable of extraordinary things together.

Experiences such as burnout are real difficulties, but you don't need to suffer alone. If you are struggling and want professional support, speak to your GP or contact your local NHS IAPT service.

Holly-Anna Petersen has been a member of Christian Climate Action since its inception. She has an MSc Psychology, a PGCert in LI Cognitive Behavioural Therapy and works as a mental health practitioner in the NHS.

17

The making of an activist

RUTH JARMAN

2.30pm. Peace at last. The cell door closes loudly and I am on my own with a Bible, Dietrich Bonhoeffer's book *The Cost of Discipleship*, a novel, a pencil, a pen, a notebook, two leaflets given to me by the policeman who checked me in, and a cup of tea.

I make myself at home by taking off my boots, sitting down on the blue plastic mattress and arranging the matching plastic pillow against the wall. The temperature is about right, though I will ask for a blanket soon. The walls remind me of a public convenience, thankfully without the smell. I have an en-suite toilet and now a room service of more tea, water and beans and wedges just served to me by a motherly policeman. I am aware that as a middle class white person, my experience of arrest is not universal, but I have hours ahead of me to write, read, pray. No phone, no emails, no washing up, nothing that needs fixing, tidying, sorting. Ironically, I feel a real sense of freedom.

I didn't use to be an activist. Until my thirties even signing a petition was outside my comfort zone. I kept my environmentalism at a respectable distance, donating regularly to WWF and Greenpeace, trusting them to do what was required to keep the earth in good enough nick.

A turning point came when I was walking down the street in Santa Cruz, California where I lived at the time. The first verse of the Bible: "In the beginning God created the heavens and the earth" came to mind, and I suddenly realized the connection between my Christian faith and caring for the earth. What does it mean to be a disciple if the ordinary way of living is trashing what God has made? From that point I started signing petitions, but I guess they were just 'gateway drugs' to lobbying, marching and finally civil disobedience, when the more respectable activities failed to provide the satisfaction of preventing climate breakdown.

I was arrested today at the Department of Business, Energy and Industrial Strategy, alongside fellow activists from Christian Climate Action and Extinction Rebellion. In the police interview they showed a video of me spray painting the XR symbol on the glass wall of the lobby.

"Is that you?"

"No comment."

"Who owns that glass?"

"No comment."

"Did you have permission to spray paint that glass?'

"No comment."

I felt uncomfortable. I am the sort of person who usually falls over backwards to be polite and respectful, especially to people in public service. However, the legal advice given to activists is that 'no comment' interviews are the best way to go and the police are used to it. The whole business of civil disobedience is controversial, especially for Christians. After all, the Apostle Paul says in Romans that we should submit to the authorities. Since he probably wrote this from prison, we need to consider what he actually meant. That obeying the law is all good and proper, but it must come second to obeying God. And when the laws of the land are actively destroying what God has made, perhaps it becomes a Christian responsibility to disobey the law; and divine obedience becomes civil disobedience. History has shown that at times it has been right to break the law to achieve justice. This is bigger than justice, it is about survival.

More fundamentally, I'm not sure I believe I was breaking the law. It is lawful to throw a brick through a window of a burning building to save someone inside. It is lawful to take action to prevent a greater crime. Our government's policies make it criminally negligent for failing to protect life. If my actions expose this truth, and help it to repent and do what it takes to protect the future, they will have been lawful.

Something has to change. The only thing that humanity needs to do to destroy God's beautiful creation, and human civilization with it, is to carry on going about our ordinary law-abiding business as usual. Conventional advocacy has not worked. In June 2017, thousands of people converged on Westminster for the biggest ever mass lobby of Parliament on climate. MPs met us, listened to us, yet that very evening Amber Rudd announced a cut in the subsidies to wind farms, the cheapest form of clean energy, the no-brainer in emissions reductions.

When I write to my MP he always replies to say that Britain is making progress, and is a leader in reducing emissions, but we increase subsidies to fossil fuels and undermine renewable energy. There are currently no policies, targets or even pledges that take us anywhere near where we need to go. We can't just nudge things along incrementally. There has to be a complete reformation of the global economy. So while petitions and lobbying should by all means continue, we need to speak louder. One way to speak louder is to step outside legal boundaries.

The first time I was arrested was at Kingsnorth in 2008, where there were plans to build a new coal-fired power station. We sat down in front of a gate, prayed and sang. The police asked us to move and we just carried on singing. Subsequent to our protest, plans for the power station were cancelled and there has been no new coal power in Britain since. We succeeded.

Will these protests be successful too? Can the global economy be turned around, and the worst of the climate crisis be averted? I don't know. We know that in the past, mass civil disobedience has changed societies dramatically. The theory is that if enough of us rebel against the current destructive system, perhaps, just perhaps we could change the course of history. God willing.

The way I see it, without Extinction Rebellion, or a miracle (or perhaps they are the same thing), I don't see how there can be hope for a safe future. With prayer and rebellion, hope is conceivable.

In the police van coming here, handcuffed, driving past the 'Tell the Truth' umbrellas of the protest at the main entrance of the Business Department, our movement looked very small, very insignificant. Yet "for us who believe" we have an "incomparably great power", a power mighty enough to raise Christ from the dead (Eph 1.19).

Today, we are Christ's body. We are his hands and feet in the world. Our head is Christ himself. And as Paul wrote – from a prison cell – the Christ we worship and serve, and in whose name we act, is "far above all rule and authority, power and dominion... in the present age and the one to come." (Eph 1.21)

Ruth Jarman is a founding member of Christian Climate Action, and a co-founder of Operation Noah. She has been arrested multiple times, in this case for an action at the Business Department in 2018.

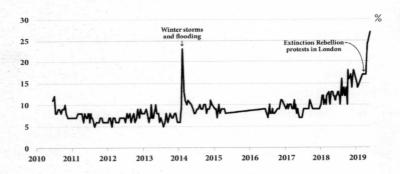

Concern about the environment at highest levels on record.
Which do you think are the most important issues facing the country at this time? Please tick up to three.
% saying "the environment".
Credit: YouGov

Ruth Jarman protests outside Downing Street, 2018.

Success stories

Extinction Rebellion protests have WORKED as MPs
succumb to calls for change
Daily Express, 23.4.19

UK Parliament declares a climate emergency
BBC News, 1.5.19

MPs bow to Extinction Rebellion demand, as they send
out invitations to climate change citizens' assembly
The Telegraph, 1.11.19

Fracking banned in UK after government U-turn
The Guardian, 2.11.19

Climate more important than economy to voters
The Independent, 18.11.19

'Climate Emergency' is declared 2019 'word of the year'
Daily Mail, 21.11.19

EU Parliament declares 'climate emergency'
Deutsche Welle, 29.11.19

18

Beyond hope and despair: journeying together through climate grief

PAUL BODENHAM

The water is wide, I cannot get o'er
Neither have I wings to fly
Give me a boat that can carry two
And both shall row, my love and I
    O Waly Waly (traditional folksong)

This is a hard time to hope. It is hard to know what hope is any more. If you are grieving and fearful for the future of the Earth, you are not alone – and because our grief is a collective experience, we can count on the consolation of others who are grieving too.

However we interpret the unfolding experience of loss, perhaps the best service we can to offer each other is to accompany each other along the way. To begin with, we shall need help from others to make sense of our bewilderment and anxiety; in due course we shall find we have more to give our fellow travellers than we thought.

According to the climate psychologist Renee Lertzman, attention to our grief is not only good for us; it helps us to advance our cause:

The more we can acknowledge openly and explicitly how we're feeling about what's going on, the more we can quickly free up a lot of that energy to be strategic, creative – all of the capacities we're needing to unleash right now.

Frontiers of courage

This is a time between worlds. From us who live in it, it demands new skill and new discourse. We need leaders of mourning, pioneers of courageous unknowing. We need makers of meaning for an age of loss. So let's explore the strange new landscape which opens up when we have the courage to let climate breakdown be the making of us.

Thomas Merton, the monk and existential explorer of the 1960s, speaks for those of us called today to the edges of the thinkable, beyond where the human mind has dared, or needed, to go:

Night is our diocese and silence is our ministry
Poverty our charity and helplessness our tongue-tied
 sermon.
Beyond the scope of sight or sound we dwell upon
 the air
Seeking the world's gain in an unthinkable
 experience.
We are exiles in the far end of solitude, living as
 listeners
With hearts attending to the skies we cannot
 understand:

Waiting upon the first far drums of Christ the
   Conqueror,
Planted like sentinels upon the world's frontier.
   The Quickening of John the Baptist, Thomas Merton

I identify six frontiers between the world we know and the world we do not. We have to remain at all of them, and our churches are called to these frontiers too, to carve out spaces where people of grief and goodwill can 'seek the world's gain in an unthinkable experience'.

1. To help each other tell the truth

Guarding safe space for ecological distress, not just for ourselves, but for society at large.

If we face the fear that arises from what we learn, we can begin to disarm it, and look more squarely at the truth that prompts it.

This bias to the unthinkable contrasts with traditional climate campaigns, which appeal to supporters with upbeat messages. Narratives of hope increasingly stretch credibility, and ride roughshod over the inner desolation of campaigners who have learnt too much. The very notion of the 'campaign ask' is now at risk – whether it's 1.5°C, 350 parts per million of atmospheric carbon dioxide, or even zero carbon by 2025, and yet where else are we to draw the line?

With a small group of people we trust, we can find courage to share feelings in which we thought we were alone. Their gentle questions give us permission to say what previously could not be said. Their affirmative reactions reassure us that we are not going mad. The sky will not fall in if we take that first step in confidence.

2. To find language for lament

By word and ritual, enabling climate grief to be articulated in the public realm.

The consumer economy is playing a trick on us. It tells us we are free to make choices according to our self-interest, and those choices make us who we are. In this impoverished model of humanity there is little room for the other: we are alone.

The system can ill-afford to let people connect and make common cause. It has conspired to prevent us finding words with which to speak the truth for our time. That's how the machine perpetuates itself.

The practice of lament subverts all that. In lamentation we don't just think the unthinkable: we speak the unspeakable. We don't just speak for ourselves: we speak for and with each other, the Earth and God.

3. To befriend impermanence and mortality

Confronting our fear of death, including that of our offspring, and the prospects for our civilization.

Since the dawn of time humanity has wrestled with the reality of death, and our disinclination to think about it. Some of our deepest wisdom has emerged from that struggle.

Climate breakdown brings a new focus to our encounter with mortality. It makes the prospect of death more immediate and more communal. Some causes of death may become more likely than in our parents' time. Yet the prospect of death itself is of course unchanged – no more likely and no less.

Many of us have lost loved ones of our own. These bereavements can be precedents which lend perspective to an uncertain future. They help us to comprehend our own inevitable personal extinction, and the extinction of species.

4. To find wisdom for the future

From traditions of self-transcendence, opening up new pathways of understanding and agency.

Traditions of ancient wisdom have been tested by centuries of bargaining with the Ultimate, and they help us live into the future. By sinking roots into them through a faith or spiritual practice we are better able to face uncertainty. Faith sets us free from the future and whatever it may hold, in order that we may be free for that same uncertain future.

In this time of contingency, there is a every risk that "hope would be hope for the wrong thing", as T.S. Eliot put it in *East Coker*. Rather, he says, "the faith and the love and the hope are all in the waiting". As saints and sages remind us, we must surrender the ego and its ambitions, sometimes even the urge to know, and to enter the Dark Night and the cloud of unknowing.

We do well to draw deeply on the tradition that lies closest to our reach. Some of us have been particularly nourished by the Christian tradition. As we enter this ecological Passion, you might find a new urgency to follow Jesus through the great arc of his incarnation, death and resurrection. We were baptized into that paschal journey, and it is as though we are being baptized again.

5. To love, reconcile and resist

Protesting the best of humanity, whatever the cost.

It is often said that the best antidote to climate despair is action. It's largely true, but let's not get busy simply to distract from our feelings. As Renee Lertzman counsels us, we need to listen to what our hearts tell us, and so does the Earth. That compassion, however heavy it weighs on us, is true prayer.

It is a paradox of prayer that the more resolutely we wait, and the wider we open our hearts, the sooner we can act, and the deeper our action's authenticity. It is another paradox that, as it did with Jesus, prayer trains us simultaneously in two contrary motions: reconciliation and resistance. As insecurity, competition and conflict continue to rise, we can expect to see the worst of humanity acting out. Yet as we pray, and live out our prayer, we can enable the kingdom of God to break through to make all things new – or if not all, at least enough to redeem the whole.

6. To hold open the door to hope

Applying ourselves to grow resolute and generous in the face of potential collapse.

We act together not just for our own sake, but for love of the Earth. Whatever the future holds, much will be lost. In fact the most vulnerable on the Earth are already losing everything.

Yet much can still be saved – and the quicker we act the more can be saved. In these last days of hope, we are among those who keep a foot in the door against the pressure to despair. Our courage, our honesty, our defiance of despair make futures possible. Let's live up to what Jesus called us – salt for the Earth.

Understanding grief

There have been a number of attempts to characterize the process of grief in the experience either of those who are dying or those who are left to mourn them.

In these models death is the defining event on a timeline. Climate change is different: it will be an immersive lifelong experience from which there will be no moving on. Even our engagement with climate grief will change over time, and psychological models can help us notice those changes.

Perhaps the best-known model of the stages of grief was developed by the pioneer of palliative care, Elizabeth Kübler-Ross. She proposed five stages:

1. Denial, a phenomenon we are only too familiar with, sometimes even in ourselves.
2. Anger, directed perhaps to governments or high-consuming individuals.
3. Bargaining, an activity which has occupied countless hours in global summits or local initiative, and with little more success than Kübler-Ross's patients.
4. Depression, when the prospects of mortality or collapse finally become ineluctable.
5. Acceptance, as the patient begins to prepare for the inevitable. On the evidence of Extinction Rebellion it is in this place of calm that judgments and choices can best be made.

There are alternative models of the grief process, such as Colin Murray Parkes's model based in loss and attachment. The UK based Climate Psychology Alliance commends a third, by William Worden. Rather than a linear process, this conceives of grief as a series of tasks which can either be embraced or rejected, and the sequence of them is less significant.

Embracing the tasks of grief	Rejecting the tasks of grief
Accepting the reality of the loss, first intellectually and then emotionally.	Denial of: the facts of the loss; the meaning of the loss; the irreversibility of the loss.
Working through the painful emotions of grief (despair, fear, guilt, anger, shame, sadness, yearning, disorganization).	Shutting off all emotion, idealising what is lost, bargaining, numbing the pain through alcohol, drugs or manic activity.
Adjusting to the new environment, acquiring new skills, developing a new sense of self.	Not adapting, becoming helpless, bitter, angry, depressed, withdrawing.
Finding a place for what has been lost, reinvesting emotional energy.	Refusing to love, turning away from life.

Making mourning

All societies have symbols, norms and rituals to help people absorb the impact of loss, individually and as a culture: what we call 'mourning'. In the UK over recent decades we've been forgetting the arts of mourning, but their time is coming back.

Extinction Rebellion has been particularly resourceful in the expression of grief, with grief circles for its members, and sober processions of ashen-faced 'Red Rebels' to honour what is lost. Borrowed Time, a project of Green Christian, is seeking to help Britain's churches to offer spaces of grieving and patterns for mourning to society at large, so that they can become centres of regenerative pastoral care which can unlock climate action.

There is plenty more room for shared practices of mourning: solemn commemorations such as the International Day for the Remembrance of Lost Species; fasts, retreats and novenas; shared identity in clothing or worn symbols; vocal practices of chanting, recitation or keening; pilgrimages for repenting and remembering.

At the moment, ours is a 'disenfranchized' grief – one that is denied public acknowledgment because of taboo or disapproval. Yet if we apply ourselves to practices like these, and open them up to others, in time ecological grief will gain the enfranchisement it desperately needs. The right to ecological mourning will take its place in the public conscience, through a similar process of social awakening as we have had in recent years to the safeguarding of children, the protection of minorities, or, to take one simple example, the right to parental leave.

Norms of ecological mourning were not there for us, but we can help to build them for others. Grief is our greatest gift in this emergency. Let's invite every citizen to join us for the journey. To ford these deep waters is to reprise the archetypal journey into which every Christian is baptized. We have been here before. It was the making of us then. It will be the making of us now.

Paul Bodenham is a trustee and former chair of Green Christian, and co-founder of Operation Noah and Borrowed Time.

19

Dear Earth

SATYA ROBYN

Dear Earth,

I sobbed as they arrested me.

I wasn't upset about my criminal record, the blank hours ahead of me in a cell, court, the fine.

I didn't feel unsafe. As the police carried me away, they checked four times if they were hurting me, if I was okay.

I wasn't ashamed, after a lifetime of being a good girl, of not inconveniencing anyone, of doing as I'm told.

I sobbed, beloved Earth, because the grief I felt for you suddenly rose up and crushed me.

I knew that 600 of us had already been through these cells, and we were hardly appearing on our national television.

I saw the whisper of my voice up against the airplane roar of those who have unimaginable power.

I recognized the system's denial about the gravity of your prognosis, as I have lived under a thick protective crust for decades.

The grief pushed its way through me, and it left me clean.

I am lucky to be here, dear Earth, as I write with the red biro a smiling officer brought me.

Later they'll bring me food, and I'll go home to my extravagant privilege.

Others are failing to coax crops from impoverished soil. Others have had their homes violently flattened. Others are watching the ice caps melt, drip by deathly drip, and they don't know what to do.

I know what to do.

I vow to witness your vast suffering, darling Earth, and pray for your coast dwellers, intricate coral reefs and nightingales.

I vow to meet the razor-sharp protection of others with peace in my heart.

I vow to do what you call me to do.

I am so inadequate, dear Earth, and I contain the same greed, hate and delusion that is strangling you.

I am asking for your forgiveness, with my sobs, and with this red pen.

Satya Robyn is a psychotherapist, a writer and a Buddhist priest, and a coordinator of XR Buddhists.

20

Standing with the crucified

MARTIN NEWELL

When we take direct action, there is always something that has to change – an injustice or oppression that we want to stop. Yet there is more to direct action than presenting demands. As Mother Theresa, Dorothy Day and various others have said, the primary thing is about being faithful, not successful. We don't only act because we think we can win. We act because it's the right thing to do. We act to bear witness to the truth, to witness to our faith in God, and to stand with those who are suffering. We act in solidarity.

Solidarity is a multifaceted idea. Part of it is to do with privilege, which is a debate within Extinction Rebellion. For those who are relatively privileged, solidarity is a way to cash in on that privilege and use it to support the cause of those who are already suffering because of the climate emergency.

There is more to it than supporting a cause though, because solidarity is about entering into that cause, and into the suffering. We try to see the world and live our lives from the perspective of those who are crucified today, and with the crucified earth. How do we see through the eyes of those who are already suffering? Solidarity attempts to enter into experience, their perspective, their eyes – and to see faith through those eyes too.

Jesus talked about feeding the hungry and clothing the naked, and how "whatever you did... you did for me" (Matt 25.40). I don't think he was just making a moral statement. I think he was making a spiritual or existential statement about himself. When people help those who are suffering or oppressed, Jesus really does feel it as if it is being done to him. Equally, those who harm others, it is as if it is being done to Jesus. That's a statement of what I would call existential solidarity, and something we should all try to enter into – a deeper compassion and empathy than simply supporting a cause. Solidarity is not theoretical, it's something we feel deeply inside ourselves.

In the Passionist tradition, which I am part of, we talk about standing with the crucified. The heritage of the phrase comes from liberation theology, and that's been developed in Passionist thought to talk about the crucified people of today. As a Passionist said to me many years ago, it's not that we say people are crucified today because Jesus was crucified. It's more that Jesus was crucified because he was in solidarity with those who were crucified in his time – and we are called to follow that Jesus.

To be crucified means to be suffering, to be oppressed, exploited, to be a victim of violence, to be degraded and tortured, to have your life taken from you, to be killed. All of those things are true of the earth, of God's creation. As Pope Francis has been reminding us recently, the Earth is our mother and our sister and our neighbour. In *Laudato Si* he describes the earth as being among the most abused of our neighbours. It makes sense then to say that the earth too is being crucified, the earth on which we depend. Creation is not an inanimate object. It is something that has life which has been given by God, and that also needs to be defended, as anyone being crucified should be defended.

We see that everything is connected in the gospels: when Jesus is crucified, the earth reacts. There is thunder and lightning, and it goes dark. We may see that as symbolic language, but it signifies how it's all one reality. What happens to us affects the earth, and what happens to the earth affects us too.

Another word for that one-ness is communion, which is also related to solidarity. 'Com-union' means 'union with'. We're all one. That's the reality of our faith and of life. The communion between us and the earth and God has been ruptured by sin, and we are trying to rediscover that communion and act on it. Perhaps that is one of the reasons why celebrating the Eucharist during Extinction Rebellion actions has been so powerful for those taking part.

If you think back to the last supper in the gospels, Jesus basically held a clandestine meal – there are these coded instructions to the disciples to enter the city, look for a man carrying a jar of water and follow him. It's planned in secret because Jesus has just followed some powerful street theatre on Palm Sunday with serious direct action at the temple, and the authorities are looking for him. The last supper happens between Jesus' actions at the temple, and hours before he is arrested and taken to court, sentenced and executed by the state. A protest setting takes the Eucharist back to its original context, to break bread together reminds us what that was about and brings it to life.

Often faith and prayer and the Bible come to life in new ways in protest contexts, because so much of the Bible is written from the perspective of people who are oppressed, who are persecuted, who are standing up for what is right – whether that's the prophets, Jesus or the early Christians. Some of them had to stand up to the authorities and got in trouble, got arrested and jailed. The Psalms in particular are something I'll turn to in a police cell or even on a prison bunk. Our faith comes from the marginalized and oppressed Jewish people, living under occupation or in exile. These are the people who wrote the Bible. This is where it comes from and where it makes sense. It all comes alive when you start to experience faith from that perspective.

Solidarity also protects us from thinking we have all the answers. People often accuse protestors of self-righteousness, and we can easily slip into that. To see our action as an act of solidarity with those who are already suffering can guard against setting ourselves up over and against people.

I try to express this during protests. For example, at a recent action at the Docklands Light Railway, we tried to go into it in a spirit of repentance for our complicity in the sins against the earth and against the poor. That's why I was kneeling down on the roof of the train, to symbolize the penance and the humility of that action. It's not like we're the ones who have got it all sorted and everyone should do what we do. Yet somebody's got to do something, and we confess our complicity as we do it.

The first time I got arrested was with six East Timorese refugees. We went onto this British Aerospace base, where we dug a small grave for a symbolic child's coffin. British Aerospace were exporting their warplanes to Indonesia at the time, and they were used in East Timor, where these guys were from. Spending 24 hours in a cell with my friend from East Timor, whose family has been killed by the Indonesians, was a very humbling and powerful experience of solidarity. This was their struggle, and we could only help to bear witness. Some years later East Timor became an independent country. Our action was obviously only a tiny part of a huge movement that achieved that, and ours wasn't even an action aimed at Timorese independence, nevertheless we felt part of that.

As we act with Extinction Rebellion, we press for change. We also stand with all those suffering from the climate emergency, viewing the world with their eyes, and acknowledging our part in their suffering. That perspective keeps us grounded, keeps us humble, and keeps us on the streets.

Fr Martin Newell cp is a Passionist Catholic priest, a founder of Christian Climate Action and the London Catholic Worker.

21

Climate eldership

PHIL KINGSTON

My involvement with non-violent direct action goes back about 17 years. I had been involved with environmental concerns before, and I had long been concerned about poverty in poorer countries, campaigning with CAFOD. Then I became more interested in peace, the arms trade and nuclear weapons at Faslane. My first action was at a factory in Wales that was producing munitions. With another man I sat in the road so that the people coming in on the afternoon shift couldn't get into work. That was my first involvement with breaking the law.

A few years later, just before we went to war with Iraq, I was arrested a few times in London. Now my actions are almost entirely in relation to care of the earth and its ecosystems. That came about with the arrival of my grandchildren. The eldest is now 20. Concern for them and their generation has become my most immediate motive, alongside concern for the poorest, and for God's earth.

Thanks to my involvement in Extinction Rebellion, I have a notoriety status that I never imagined. The action that got the most publicity was stopping the train at Canary Wharf in April 2019. That went viral. I get recognized now, and I've been taken unawares by that. For example, one time when I wasn't supposed to be within the M25, and I was. I was at a protest, and somebody called out my name. I thought it was one of my colleagues and I turned around, and it was a policewoman. So she had me. I had been wearing an orange jacket which is rather recognizable. I don't wear that now in London.

Other times younger people come up to me at protests and say "thank you for being here." I find this very moving, and also very interesting. When I've had the chance to talk with them a bit, I can see that we older rebels are giving them support and encouragement: for some it's almost permission. For others it's something much deeper than that. Many people have not experienced, or are not currently experiencing, the support of elders, either within the family or elsewhere. Perhaps there's something biological in this, going back deep into our culture and our past, but it seems that the presence and support of elders is something that is really longed for.

Elders also have unique opportunities, and I began to see this about nine years ago. With others in the Bristol area, we set up a group called Grandparents for a Safe Earth. We had the same commitment to civil disobedience and nonviolent direct action that Extinction Rebellion has now. What we quickly discovered was that the police didn't want to arrest us. We were arrested only once and then de-arrested when they took us out of a bank we were occupying. That opened up a very interesting topic around the role of elders, because I think there is a certain amount of respect for us; and when working as an affinity group, we are accorded an unusual level of social power.

I see this with the police. I'm aware that my story – speaking as a white, older, middle class and rather recognized

Phil Kingston under arrest
outside HM Treasury, October 2019.

man – is different from many others. So I speak for myself. My experience with the police is that they have been very respectful, and concerned, and that's to do with me being older. Being arrested is very different as an older person. I am very rarely handcuffed. One policeman said as we were waiting for the van, "You're not going to do a runner, are you!?"

Perhaps most importantly, the police are often willing to speak. They know that we within XR are not a danger to them. Whenever I feel it's appropriate, when I've just been arrested or when in the police station, I engage with the police and tell them I'm doing this because I'm a grandparent. I ask if they have children, and they respond to that in ways that I think they wouldn't normally. One senior police officer, when I said that, told me "We have 11 years to turn this around, and after that it's a disaster for my children." That's perhaps been the most clear statement that we have a common cause.

Another time I was in a police station that had been opened up specially to deal with XR, and there were a large number of us kept in overnight. In the morning one of the older officers who was going off duty came to my cell and wished me all the best, and he said "You people are the best customers we've ever had". I knew what he meant – we were no trouble. We did our best to acknowledge that they've got their job to do and we've got ours. There was a good level of humour and respect.

The differences for older activists go beyond arrest of course. Younger people have so many concerns – how to get going financially, how to build a career. I have met some in

XR who have put a lot of things on hold to do this work, and they have my huge admiration. I am free of all of that. That's a big aspect of being an elder. We are often much more free than most people to do many actions, and the consequences really don't matter in the same way.

Eldership has become something of a responsibility too. Our actions in Canary Wharf were seen by many at the time as something positive and meaningful. It focused on a major financial centre in the UK, linking that to the funding of climate emissions and environmental destruction. That was not the case with the second train action I was involved in at Shadwell tube station in October 2019. There was a lot of verbal violence, shouting, and distress from people there. That experience of being immediately faced with people who were distressed by something I was causing, that led to a lot of rethinking.

Non-violence is one of those terms that is phrased in the negative, and if I was to try and make it a positive I would say compassionate behaviour. I do feel a lot of compassion for those people now. I wasn't able to connect with them, anything like enough, in the heat of that 20 minutes or so at Shadwell Station. Since then I've made a decision that I will not take actions that have a random, unpredictable and unknowable impact on others. I would rather take actions that are very focused, geographically, so that I am present to take responsibility for what I am doing. This is a conversation I have since been involved in with colleagues, and perhaps there is a blessing attached to that experience of an action going wrong, because it has opened up new conversations that will make us more compassionate in future. That would be very positive for our future actions and our relationship with the public.

I'm grateful that, because of my higher profile, I had the opportunity to speak about the Shadwell action with the press a few days after I got home. I was able to say that I understood the community anger. These were ordinary people doing their best for their families. I regretted adding to their difficulties in life. I was able to say that I will not be taking that kind of action again. I'm grateful for that opportunity to speak my truth, because it in some way made reparations and showed how we move forwards and do better. That is part of eldership too.

I'm 83, and my competencies are dropping. There are changes in life that are happening and they're not going away. Physically and mentally, I'm slowing down. I don't deal with complexity as well as I once did. I don't deal with pressure as well. This is the real me, that is not captured in the snippets of truth that circulate in online videos. There are downsides to being older, but as I have described, so many privileges and opportunities as well. And so I find myself speaking more and more about elders, and to elders, to encourage them to do whatever they are called to do.

Everybody, whatever age, has something they can contribute.

Phil Kingston is an activist and founding member of Christian Climate Action.

'Don't ever apologize for crying for the trees burning in the Amazon or over the waters polluted from mines in the Rockies.

Don't apologize for the sorrow, grief, and rage you feel.

It is a sign of your humanity and your maturity.

It is a measure of your open heart, and as your heart breaks open there will be room for the world to heal.'

Joanna Macy

22

In court – exactly where I'm supposed to be

SAMANTHA LINDO

The coach was booked for 4.15am from Bristol. We were legally required to be at City of London Magistrates Court by 9.30am. But having packed my bag, made my lunch, and (apparently) set my alarm for an early start, I opened my eyes to sunlight.

"It's five to seven!" I gasped, almost pushing my husband out of bed.

Five minutes later we were pulling onto the motorway; me still in my nighty, clutching my toothbrush and belongings.

I was settled in my head that I would plead guilty. I had discussed my options at great length with my lawyer from Hodge Jones Allen, a firm specializing in representing activists. If I pleaded guilty, but went into the court unrepresented, I would have the chance to read a statement and have my voice heard without having to go forward with a trial at this time; it felt like the right thing for me to do.

I arrived just in time to a foreboding building which I was thankfully distracted from by the welcoming sight of people waiting in support, smiling, greeting me, laden with XR badges.

It was busy. Everyone seemed to be in the same boat, nervously chatting and sharing stories. People were handing out fruit and chocolate and someone went out for coffees – and then suddenly I was called –

"Samantha Lindo, guilty plea, unrepresented".

The next thing I knew I was sitting on the back row of three tables in front of the judge. A whole row of people behind me – legal advisers, other defendants and media volunteers. A lady was just finishing off her not-guilty plea, discussing the details of the court date with the judge.

I sat with a dry mouth. Was this really happening? Not just the fact that I was there, but that the earth and us as humanity face this reality. I'd always known it, felt it, lamented it quietly, but this outward experience of the state of things sent a new wave of crushing disbelief and grief through my bones. Perhaps, like all paths in Jesus' life were leading to the cross, all the paths in my life had been leading to this. The cost of telling the truth and acting accordingly. The judge invited me to take a seat at the defendant's table.

I was asked my name, address, date of birth and plea. A man adjacent to me, who I soon realized was the prosecutor on behalf of the police, started to speak, recalling all the details of my arrest.

I had joined the protest straight from an Easter morning service at St John's Waterloo. The female curate (who had been hosting activists in the crypt all week) had given a moving sermon in support of XR, celebrating the radical love of Mary Magdalene, the devoted woman who watched and waited with Jesus until the end, unafraid of facing the parts of reality that many would rather turn away from. We witnessed the baptism of a small, innocent child, a child that had no knowledge of the situation we face, yet who will reap the consequences of our choices now.

As we arrived onto the occupied bridge, an indigenous woman from Brazil, who held the crowd in the palm of her hand with her words and songs, told us how when they march to protect the last parts of the Amazon, their home and indeed, the very lungs of the earth, they are met not just with the police, but with the military. The privilege of our democracy and our right to protest safely shone in my mind and I felt compelled to use it.

At that point, an army of police officers descended with orders to move us to the legal protesting site at Marble Arch. At this point, I looked around to what seemed like a sea of women (men too, but for some reason around me, predominantly women). Later in the police van a female officer commented on this. I replied instinctively: "Well, one hundred years ago, we would have been getting arrested for the vote."

She agreed and nodded. Myself, the officer and the two other ladies who I'd been arrested with shared a smile and a quiet moment of solidarity.

In the court room the prosecutor read the words of my arresting officer which described in detail how he had previously warned a group of women, sitting in the road, singing, of the Section 14 imposed on the protest. On returning to them, he continued, they "Continued singing and started applying lipstick".

The courtroom erupted with giggles and I tried my best to conceal my smile from the judge. Then the mood changed. He looked at me.

"Ms Lindo, do you have anything to say at this point?

I stood, my legs shaking.

"Yes, your honour". I started to speak. "I have pleaded guilty today because I did understand the Section 14 imposed on the protest on Waterloo Bridge on Easter Sunday and I chose to remain present."

I continued: "The officers present were clear, kind and professional in communicating this to me, as they were all the way through my detention, something I appreciated through what was the somewhat daunting experience of my first arrest."

The judge nodded.

"As a teacher of young people suffering with their mental health, working in the public sector, I understand first-hand the stretched resources that the police force are faced with and I am sorry my arrest took up some of these precious resources."

This was well received, I noted. "I am also sorry if any of those resources were diverted from supporting young people, something that was constantly on my mind throughout my night in the cell. But, more than anything…"

I took a deep breath.

"I am sorry that it has come to this. That I feel I have to get myself arrested in order to get my voice heard."

The words spilled out of me now, no looking at the page.

"I have campaigned, marched and legally protested for the whole of my adult life. We have been ignored. This is an emergency. Unprecedented."

My voice broke now and you could feel the weight of the room.

"It has got to the point where I can no longer look the young people I work with in the eye and tell them I believe there is hope for their future. Or, as a recently married woman in her thirties, when everyone is expecting me to have a child, to even consider this a reasonable or indeed moral thing to do, or retain any sense of emotional and psychological congruence within myself without acting in accordance with this emergency."

I took a breath.

"Nor as someone with a Christian faith, who is called to stand up for the vulnerable and tell the truth."

My eyes met the judge's as I drew my thoughts into a close.

"So, Your Honour, if I am guilty of acting in accordance with this emergency – to protect myself, my future child and indeed all of us… then so be it. But, if I am guilty, I would like to ask you and all here present, how much more guilty is the government, that holds power in this country, for not acting, not telling the truth and not doing its ultimate job of protecting our lives, the lives of our children and indeed all life on this planet? Thank you."

There was a pause. The judge, broke the silence.

"Well, Ms Lindo," tripping over his words slightly. "I have to say, you have expressed yourself… most eloquently."

He looked moved, and possibly in agreement.

"I am giving you a six month conditional discharge and a fixed penalty of £85 plus victim charge of £20 totalling £105."

We discussed how I would pay this, all the while standing, my legs feeling like jelly.

As I turned to leave the room, the court erupted with a huge round of applause. I hadn't expected that. Everyone who was at the back stood, it was wonderful to walk past the emotional, smiling faces of those who had been there with me to share this moment.

I floated out into the waiting room to be congratulated and thanked by an array of people, and to my surprise, a lady from CCA was there as arrestee support collecting all my details, ready to give me a big hug.

I knew in that moment that this was one of the most meaningful moments of my life. I felt I was exactly where I was meant to be, doing what I was meant to be doing. Acting in non-violence. Demanding justice. Telling the truth.

It felt not only wonderful, it felt divine.

Samantha Lindo is a Bristol based singer-songwriter of mixed Jamaican heritage.

23

Wisdom from the whirlwind – a call to humility and joy

BILL MCKIBBEN

The book of Job describes its titular character struggling with the orthodoxy of his day – with the 'obvious' notion that God dispensed prosperity to the good and punishment to the wicked. We 21st-century Westerners are beginning to struggle with an orthodoxy of our own – the central economic and social idea that more is better, that growth is necessary. These two orthodoxies are similar in that an examination of the facts (Job's innocence, our own climate emergency) call them into serious question.

They are similar in another way as well: both stem from the assumption that human beings are and should be at the centre of everything. They are out-growths of the human-centred or anthropocentric understanding that has dominated modern human culture. Job's friends believed that God's calculus was their own, that he made his plans with regard to our very human ideas of justice and fair play. We believe at some intuitive level that it is all right to use everything that surrounds us for our own benefit – that creation matters because it is of use to us.

Some commentators trace this notion to the beginning of the Bible, to the Genesis idea of humans holding dominion over the earth. The validity of this reading is up for debate, but I am more interested in the reply that God provides to this anthropocentrism when he speaks from the whirlwind at the end of the book of Job. It is, I think, a shatteringly radical answer, one that undercuts the orthodoxies that entwine us.

The first disconcerting aspect of the long speech from the whirlwind is its tone. This is a deeply sarcastic God speaking: "Where were you when I planned the earth? Tell me, if you are so wise."

Even more notable, however, is the setting. God is describing a world without people – a world that existed long before people, and that seems to have its own independent meaning. Most of the action takes place long before the appearance of humans, and on a scale so powerful and vast that we are small indeed in the picture of things. Almost the only reference to our species in the speech makes the point absolutely clear: "Who cuts a path for the thunderstorm and carves a road for the rain – to water the desolate wasteland, the land where no man lives?"

God seems untroubled by the notion of a place where no man lives – in fact, God says he makes it rain there even though it has no human benefit at all. God makes the wilderness blossom – a strong rebuttal to the idea that we are forever at the centre of all affairs. The first meaning, I think, of God's speech to Job is that we are a part of the whole order of creation – simply a part.

That is of course a radical idea, and yet it should not come as such a complete surprise. It is certainly not the first such hint in scripture, which begins with God's creation of everything else, and God's instructions to the living creatures to be fruitful and multiply and fill the earth. Who was Noah, after all, but the original radical conservationist, assigned to save a breeding pair of everything in the hope that they would one day again spread out across the world in their natural habitats?

My favourite vision of this balanced world comes from the familiar 104th Psalm, which of course begins with creation, using almost the same images as the voice from the whirlwind but a kinder, gentler tone. The Psalm describes the springs gushing in the valleys, watering the beasts of the field and also the wild asses and the birds of the air.

Thou dost cause the grass to grow for the cattle, and plants for man to cultivate, that he may bring forth food from the earth, and wine to gladden the heart of man, oil to make his face shine, and bread to strengthen man's heart.

So our needs are to be met, and met handsomely by the world. Yet that is only part of the story.

The trees of the Lord are also watered abundantly, the cedars of Lebanon which he planted. In them the birds build their nest; the stork has her home in the fir trees. The high mountains are for the wild goats; the rocks are a refuge for the badgers.

Not "the wild goats live in the high mountains". The high mountains are for the wild goats. Not "the badgers seek refuge in the rocks." The rocks are a refuge for the badgers.

That clear evidence has always undercut our anthropocentric bias, at least to those who would honestly examine it. However, many of us live in the cities or suburbs, and so the evidence is less obvious to us – these environments are designed with human beings at the very centre, manicured to remove the thorns and sloped to drain the swamps. Yet we do have access still to a great deal of secondhand information about the world around us. Not just the furry and cute animals, but the whole that they compose, a complex and gorgeous artefact of the mind of God that we insist on stripping down, simplifying, weakening, impoverishing.

The facts – the testimony of the psalmist, the evidence of our own eyes and ears, the stark warnings of our scientists in the face of climate breakdown – lead to the same conclusions that God draws for Job in his mighty speech. Our anthropocentric bias is swept away. The question becomes this: what will replace it?

Humility, first and foremost. That is certainly Job's reaction. If we are not, as we currently believe, the epicentre of the created world, then we need to learn to humble ourselves. Humility is usually regarded as a spiritual attribute, a state of mind. Yet it also has more practical aspects, leading us to walk more lightly on this earth, with more regard for the other life around us.

Most cultures, historically, have put something else – God or nature or some combination – at the centre. But we've put these things at the periphery. A consumer society doesn't need them to function, and it can't tolerate the limits they might impose; there's only need for people. Our culture is a natural extension, magnified a thousand times, of the culture that Job and his friends inhabited; it revolves entirely around us, leaving very little room for anything else. What we need to be figuring out, in this time of crisis, is nothing less than what the proper relationship is between people, the earth and God.

That involves the second great point of the voice from the whirlwind. That voice does not confine its remarks to the propositions "You are small and I am large", or "I am old and you are an infant". It does not speak to us purely in rational terms. It does not call us only to humility.

The voice also calls us, overwhelmingly, to joy. To immersion in the fantastic beauty and drama all around us. It does not call us to think, to categorize, to analyse, to evaluate. It calls us to be. The reason Job matters so much to me is because of the language – the earthy, juicy, crusty, wild, untamed poetry of God's great speech. The images are visual, sensual, a total contrast to the academic exercises that occupy the previous thirty chapters of Job. God simply embarks on a defiantly proud tour of the physical world, a world filled with untamed glory that reflects his own:

> Do you hunt game for the lioness and feed her ravenous cubs, when they crouch in their den, impatient, or lie in ambush in the thicket? Who finds her prey at nightfall, when her cubs are aching with hunger?

This, remember, came in a day when lions were still a real threat to people – and yet beloved of God. Provided for by

God. And it is not just the fierce that God boasts about, but the disgusting as well.

> Do you teach the vulture to soar and build his nests in the clouds? He makes his home on the mountaintop, on the unapproachable crag. He sits and scans for prey; from far off his eyes can spot it; his little ones drink its blood. Where the unburied are, he is.

Job complains that the world makes no sense and God shows him the little vultures drinking blood. That is his answer. We are beyond categories here, and into the rich, tough, gristly fabric of life.

> Who unties the wild ass and lets him wander at will? He ranges the open prairie and roams across the saltlands. He is far from the tumult of cities; he laughs at the driver's whip. He scours the hills for food, in search of anything green.

The tenderness of that last line – the hungry ass, scouring the desert hills for food and yet laughing at the order and security represented by the stable. The meaning is clear. Not only are all these things mighty and inexplicable and painful, but they are unbearably beautiful to God. They are right. They should brew in us a fierce and intoxicating joy.

It is not a storybook that we were born into, but a rich and complicated novel without any conclusion. Every page of this novel speaks of delight – not rational, painless, comfortable, easy pleasure, but delight.

I have experienced this in the natural world myself, an acceptance that convenience and comfort and ease are secondary goals at best, and sometimes very much in the way of actual experience of the world's glory. Those wild days stay with me – wet, dusty, rank, crackling, mushy days, bear days and eagle days, bee-stung and loon-sung days. This non-rational world of smells and sounds and sights, of immersion, of smallness and quietness, answers to some of our deepest yearnings.

This kind of untamed joy, or rapture, is not confined to our dealings with the natural world, of course. Friendship, love, sex, rewarding work or sacrificial giving – all can be openings into this irrational and deeply moving world. None move according to the calculations we are taught by economist and advertisers. All produce joy far deeper than any material acquisition, the joy of immersion in something outside of yourself, something larger than yourself.

The challenge before us is to figure out how to link these two callings, these two imperatives from the voice in the whirlwind – the call to humility and the call to joy. Each one, on its own, is insufficient. Yet together they are reinforcing, powerful – powerful enough, perhaps, to start changing some of the deep-seated behaviours that are driving our environmental destruction, our galloping poverty, our cultural despair. Fortunately, the two can go hand in hand.

Bill McKibben is an author, educator, environmentalist and co-founder of <350.org>.

'Those of us who witness the degraded state of the environment and the suffering that comes with it cannot afford to be complacent.

We continue to be restless.

If we really carry the burden, we are driven to action.

We cannot tire or give up.

We owe it to the present and future generations of all species to rise up and walk!'

Wangari Maathai

24

Fire is fire

VANESSA NAKATE

Fire is fire regardless of where it burns!
The Amazon is burning
Australia is burning
California is burning
Africa is also burning
The forests of the world are burning!
The fires in Africa are not different!!
They are also fires
They burn and kill!!!

The news stations are silent!
But that doesn't silence me!
They were silent three months ago!
They are silent now!
They will keep silent!
Until we speak up loud enough for the world to listen!
We will make the headlines for Africa even if you don't
    want to!!

I don't have the reports you want!
I carry the tears of people in my heart!
I carry the pain of the animals in my heart!

The Heart

That is enough for me to speak up!
People are dying
Animals are dying
We need to speak up to survive!

I will be striking for a better future
See you on the streets

Vanessa Nakate is a business student and a pioneer of the youth strike movement in Uganda.

25

A visible Christian presence

HELEN BURNETT

In October 2018, amidst increasingly alarming reports about climate change and with extreme weather conditions beginning to affect privileged white communities long after it had wreaked havoc with those most vulnerable to serious climate breakdown, a letter was written to *The Guardian* endorsing the use of non-violent civil disobedience as a necessary and reasonable response to what was increasingly referred to as the sixth mass extinction. The fact that Rowan Williams was a signatory to this letter added to my mounting feeling that words, petitions and marches were not enough.

The realization that humanity is now the single biggest factor in environmental change, and a sense of deep despair about the future of the planet and what that meant for my nearest and dearest, found me standing in Parliament Square in October for the launch of a new initiative: Extinction Rebellion. Highly visible and all alone in a clerical collar, I was surprised to be approached over and over again by people pleased to see a visible Christian presence.

I went straight from the launch to an event at Southwark Cathedral where I 'flyered' the seats in the nave during a clergy conference. I had permission from the Dean so it was hardly 'disobedience', but it felt like the beginning of something new...

Extinction Rebellion is no longer the new exciting phenomena, and how the 'rebellion' continues is not clear, but on two Saturdays in October and for a spell of ten days over Holy Week and Easter I experienced something quite extraordinary.

Extinction Rebellion is founded on the principle that the climate catastrophe we now face is the result of both a material and a spiritual crisis for humanity; that the disconnect between consumer and producer, the reduction of everything to economic value and growth compels us to respond to the crisis; and that the response must be grounded emotionally as well as practically.

XR has a flair for using symbols and ritual. As the first day of action, in which five London bridges were blocked, drew to a close, hundreds of protesters returned to Parliament Square for interfaith prayers and tree planting. The following week the celebration of peaceful protest on the bridges turned to a solemn funeral procession, including once again the use of 'the pause'. Stopping all noise and sitting or kneeling outside 10 Downing Street protestors prayed, paused, fell silent, a young man placed his hand on my shoulder. If felt as if, as one, we all 'held' our grief and our planet in our hearts and minds and whatever each of us knew as God.

I have always been drawn to both contemplation and activism and I was discovering a place where my life and my ministry felt completely integrated. By the time I arrived

Reverend Helen Burnett
October 2019

at Marble Arch on day one of the Easter action I was both terrified and determined. At home I had preached on 'crucifying creation'. I had cited Phil Kingston in sermons, and now I was a signed up member of Christian Climate Action and Extinction Rebellion.

From Palm Sunday to Easter Week I lived between my parish and Marble Arch, where the protestors gathered there created the most open, selfless and loving community I have ever experienced. This was 'Kingdom' living, and my theology and my life were fully assimilated. At times I was more frightened than I have ever been, not of arrest but of where I was being taken by events, catapulted into the limelight by virtue of a collar, called upon and applauded by people who would never dream of attending a church service. Speaking in Oxford Circus on the pink 'Tell the Truth' boat, praying as protestors were arrested, holding a whole situation in prayer – protestors and police alike, taking to the stage in Marble Arch on a daily basis to proclaim the 'Vision Statement', de-escalating a situation with the police by leading protestors in song.

Before the days of rebellion began, a huge variety of actions were planned for each day. I was part of the XR Vision Sensing team tasked with holding the 'heart centre' of the rebellion, one of only two distinctively Christian voices in a group made up mostly of Buddhists and Pagans. This was Holy Week. When asked to give a Christian perspective on non-violent action and rebellion the material was right there – the Palm Sunday Procession as protest not parade. Foot washing on Maundy Thursday would have been a weird intervention had I not been embedded in the community. As it was, by Thursday, I was known by the Marble Arch campers. My tent was on the front line of the Edgeware Road blockade, and so turning up with bowls and towels to wash the feet of rebels was entirely appropriate, received with gratitude and with interest. Grounding in the Tridium, commuting between leafy Surrey and the gritty realities of protests, never before has the liturgy felt so raw and so powerful.

We walked through the Passion to the crucifixion, and towards the active hope that Christ brings in the new creation. The horror of a creation that has been crucified – countered by a group of people living in love and seeking the risen life for all that lives and all that has been destroyed.

Where next?

A thread that runs throughout the rebellion movement is the need for grieving, for facing the truth and sitting with that truth. Again and again I have turned to the Psalms to find the language of lament and protest, again and again I was struck by the process people had gone through to arrive at the point of rebellion, again and again we shared stories of lying awake at night gripped with fear for the future. Here was a place where that fear was named and acknowledged. Returning to 'normality', it is clear that the majority cannot face that fear, that we block the reality because it is too much to bear. I think that the church could have a vital role to play in walking alongside the world as it moves inexorably towards the sixth mass extinction.

But what is it we walk with?

Are we accompanying a funeral carriage and hosting a wake, or are we midwives to something new?

I suspect we are a bit of both. I sense a shift in the vocabulary. The media have now adopted the phrase 'climate emergency' in place of 'climate change'. Extinction is spoken of, lost species are listed. We have our prophets (David Attenborough, Greta Thunberg), we have our lawgivers (Polly Higgins), we have our thinkers (Jem Bendell, Rowan Williams), we have our politicians (Caroline Lucas), we have our activists (Christian Climate Action and Extinction Rebellion), and we as Christians have a unique narrative that we can offer as humanity faces the challenges wrought by our desecration of the planet.

Above all we have the Holy Spirit, something that as a liberal and non-realist I wrestle with, but I think I would be content to say that the spirit moved in central London this April, and that is something I want to witness again. A risen life for all creation is something for which I will disobey earthly powers.

Helen Burnett is an Assistant Priest in the Caterham Team Ministry and a member of the Council of Modern Church. This article first appeared in Modern Church's newsletter Signs of the Times, *and is republished with permission.*

26

Rebel Eucharist

ADAM EARL

"Though we are many, we are one body, because we all share in one bread." I've spoken these words many times before, but this is the first time I've uttered them at the centre of an occupied Trafalgar Square, during a rave, with police slowly encroaching upon us, threatening to arrest people and dismantling a campsite. If someone previously told me I'd be in such a situation I'd have seriously doubted them, yet there I was. And not I alone, but alongside a group of fellow strangers assembled with a shared sense of displacement, desperation and purpose.

A few weeks prior to Extinction Rebellion's October 2019 actions, I saw a Christian Climate Action appeal for volunteers to help coordinative Eucharist services during the 'International Rebellion'. The climate crisis had been at the forefront of my mind since April that year, and I found myself wanting to do something to help change the catastrophic collision course our species is currently accelerating towards. I'm no priest (at least in the institutional sense), nor even a particularly textbook Christian (if such a thing exists), yet something about celebrating the Eucharist in this context seemed right to me. I volunteered to help co-ordinate these services, wondering what I was getting myself into.

After a few weeks of organising logistics (how do you source freshly-baked bread in the middle of a rebellion? Does communion wine break XR's prohibition on alcohol? What happens if the police move us on? etc.) around our day-to-day lives, we finally had a rough plan of what might transpire and how: a daily service for the first seven days of the rebellion, with the possibility of further services during the following week.

I was present at the second service. The original plan was to perform the service on Lambeth Bridge, but police re-took the bridge and by the second day had moved us on to Trafalgar Square, where CCA established camp by Nelson's Column. As people started to gather in the CCA tent for the service a commotion was heard outside: police had started to forcibly remove a key infrastructure tent that housed food for the rebellion, whilst other rebels were trying to relocate it. Someone suggested that instead of having the service in our tent we should walk out into Trafalgar Square and worship there instead, alongside the conflict between the police and the XR catering group. It seemed like the right thing to do. And out we went, chanting from the 46th Psalm "be still and know that I am God" as we walked. When we arrived, a folding table was turned into a hasty altar, and bread and wine placed on it. We stood in a circle and the priest began the service.

Dislocation abounded. This is not where Eucharist services are performed, right? The familiarity of a church, the intimacy of a friend's house or the hallowed presence of a cathedral are more fitting locales... but not here, surely? In the midst of a capital city – one of the richest and most powerful in the world – that had found itself taken over by rebels demanding a different way of being. Loud, chaotic and distinctly 'non-Christian' music blaring from a makeshift stage a few meters away. Traffic routes shut off by roadblocks composed of samba bands, a funeral hearse and people glued into bathtubs. Police marching ever closer towards the centre, with many wondering when they'd start pulling people out of the road and putting them in handcuffs. The constant thundering of helicopters circling above.

The reality of what we were doing started to hit home. Adrenaline from a day of stress was surging through my body, causing me to tremble. Earlier that day I had been engaging passers-by outside St James Park, whilst other rebels from my part of the country had erected scaffolding in the road and fixed themselves to it, shutting off the road leading to HM Treasury. I watched as people were arrested and carried away, willing to go to prison rather than stand by and do nothing as humanity continues to ally itself with forces that are pulling creation to pieces. The reality of everything was starting to catch up with me now, and I quaked. Yet this was more than just delayed shock: it was also a distinct knowing from somewhere deep inside that what we were doing in Trafalgar Square at that moment was right, and in some sort of alignment with Truth. A sense that we trod on holy ground, regardless of its outward appearance or sanitized assumptions of what holiness should look like.

I offered to do a reading during the service. I read from the Epistle of James, whose author was trying to shape an alternative community that lived in resistance to the imperial and economic norms of the Roman Empire. The final chapter of this epistle carries a stark warning to those who have spent their lives accumulating wealth at the expense of others: God is on the side of the poor, a judgment is coming, and money isn't going to save you. It's not the sort of passage you read at a wedding, or when you want to feel warm and fuzzy inside. Yet although the author's historical situation differs from ours there are broad points of similarity: we too have built a society founded on the love of gold, which exploits the poorest in our global community, and which is now facing a future reckoning that is going to be hard to spend its way out of. Food for thought, as we knelt outside a Tesco Express.

After the readings we moved on to prayers of confession. I've sometimes found the more traditional liturgy of penitence to be excessively self-loathing and too focused on individual naughtiness – such an uncompromising view of oneself doesn't engender sound mental health. Yet the climate crisis has helped me to recover a less self-oriented concept of sin, without completely negating individual responsibility. The crisis we are in is something that affects all of humanity. It's systemic and built into the very fabric of our world, and it's causing our downfall. As individuals we contribute to this systemic wrong. Despite some effort to change my individual behaviour, I continue to live a carbon-infused life and I support systems that perpetuate this present crisis. I can't help but do so, and we all need deliverance. To acknowledge our individual sin and refuse to judge others inoculates us against charges of hypocrisy as we rebel against systemic injustice. People can accuse us of owning cars or smartphones, but we're not telling individuals to adopt a way of life that we refuse to live ourselves. Instead, we're proclaiming that we are all wounded by a high carbon existence, and we need systemic change so that we as individuals can respond in kind. To fail to do this will bring judgment upon this and future generations.

As the prayers of confession concluded we shared the peace with each other: a simple act of shaking hands and wishing the other well, despite most of us being complete strangers. We prayed the Lord's Prayer in union, asking that the heavenly kingdom that Jesus spoke of might be made apparent here on earth.

As we passed around the bread and wine we immersed ourselves in a sacrament initiated by Jesus himself at the Last Supper. The Church has spent centuries theologising about the meaning and significance of the Eucharist and Jesus's death, and while metaphysical speculation has its place, I do believe we lose something when we separate the act of Communion from the narrative it arose from, and the one which we currently find ourselves embedded within. Jesus and his disciples held a sombre Passover meal, as he explained that his vision of being Israel's Messiah meant being put to death by the authorities, not overthrowing them by force. This was by no means a popular viewpoint. A crucified Messiah was unthinkable: a scandal to some and a joke to others. Yet Jesus thought that utter brokenness and dejection was the very place where God was to be found, and typified the sort of role reversal in the new kingdom that he spoke of. As we shared bread and wine with each other in

Trafalgar Square, we aligned ourselves with this unusual messianic trajectory: to challenge powers of greed, injustice and destruction, not by a superior show of strength, but by resistant, faithful witness to an alternative reality born out of sacrifice. A reality where the last shall be first, the hungry fed, where conflict has ceased, where military and financial superpowers are muzzled and their power is found to be void.

As the service ended we found that the earlier commotion had settled somewhat, with police and rebels taking up new locations for the time being. We returned to our tent, again chanting on the way. The next day six people present at the service were arrested. We hope that their sacrifice, along with approximately 1,800 other rebels arrested in London during the International Rebellion, will call forth a new way of being that our planet is crying out for.

Adam Earl is a freelance writer based in Ely, Cambridgeshire. He tweets via @adamearl.

27

Life in the shadow of death

HANNAH MALCOLM

There is a tipping point to knowledge about a dying world, where grief cannot be undone. I have reached that tipping point and cannot go back, no matter how much I try to guard myself against future exposure to the relentless cycle of bad news. I hear a bird singing and my heart drops at the conspicuous absence of its fellow vocalists. Its tiny, dark, vibrating body becomes a fluttering point of hot, resistant life in a cold sky, stark in its loneliness against the white overhead. I walk through a planted Sitka forest and project human desires and fears and griefs on each temporary line of trees, rows of timber shooting up from the sterile ground in darkness and in silence.

In the 1940s, Aldo Leopold described one of the penalties of ecological education as living "alone in a world of wounds." In 1989, Bill McKibben preferred to walk in the woods in winter, "when it is harder to tell what might be dying." Come 2019, and Mary Heglar identifies 'climate vision' as the ability to see climate projections all around you – sudden flashes of rising seas, dead people, deserted communities. Their grief expressions have all become vital mobilising responses to our crisis. Yet I have heard it said that people who work with the dying start to see death everywhere. I find myself in danger of flattening the state of the world, diagnosing even the healthy with death, those recovering with a terminal illness. Sometimes I wonder whether I can still hear and accept good news stories of species recovery, human repentance, and hopeful action, or whether the shadow of death will hang forever over every walk through the woods.

'Planetary hospice' as a model for framing environmental activism is not new. Its popular use emerged in 2014, thanks in large part to Zhiwa Woodbury's application of the five stages of grief to our current crisis. He proposed that experiencing the end of life (the 'Great Dying') in the 'Great Anthropocentric Extinction' was equivalent to being given a terminal diagnosis. Rooted in his experiences as an eco-activist attorney, long-term hospice volunteer, and 'eco-psychologist', he proposed a palliative care response which focused on alleviating suffering. His paper went viral.

The subtlety of Woodbury's argument was lost on almost everyone, the headlines claiming that he thought we were all doomed. Yet he did not intend a 'hospice' model to be taken as a diagnosis of all-encompassing death. His hospice model is consciously rooted in not knowing about the future, but in knowing that the present is precarious enough that we are moved to compassion, not despair:

> To unnecessarily descend into the kind of extreme dystopianism that sees humans snuffing out all life on planet Earth itself suffers from the kind of hubris which has landed us in this predicament to begin with. In nearly three decades of advocating for environmental

sanity, one of the most persistent foes I came up against was hubris: the idea that humans actually know what the hell we are doing and can predict how things will turn out.

Woodbury did not intend to legitimize despairing behaviour. Feelings of grief are a response to the real and are therefore meaningful as an expression of being human. They have the potential to be transformative. However, feelings of despair are not the same as grief, and their rootedness in the real or meaningful should at the very least be up for debate. Encouraging despair in others is not just debateable, but actively cruel.

How does that slip happen, from rightful grieving of what is lost to the hubris of despair? It begins when we narrow our vision for how to act in the face of disaster. If we convince ourselves that we alone know where human history is heading, we also convince ourselves that we alone know how to save it. Salvation – or even responsibility for what we consider truth telling – is too large a burden for any one person or group to bear. If we are part of a movement which we have denoted as world-saving, we are simply extending the kind of hubris that tells us we are able to fully know and therefore save/condemn more than we are able. The reality is that we cannot even save ourselves. Feeling isolated in a sea of denial, we cling desperately to our own feelings as indicators of truth, casting ourselves as the last noble stand against apocalypse in the history books. This is the first hubris.

The second hubris comes from the realization that we cannot save anything at all, or perhaps the nagging anxiety that our attempts to save things might have made things worse. All of the lights go off at once. After all, if we cannot save things, if we are not part of the solution, then we cannot imagine one. We gorge ourselves on bleak warnings from scientists and newspapers. We climb a dark mountain and cast ourselves as survivors in a grand apocalyptic drama, the only ones who could see the truth. The dark side of despair is the ugly hope that it might be vindicated, that others will see we were not wrong. It would do us good to remember that our culture is a colonising one. Hubris is in the water. We would be foolish to imagine that it would not infect even our emotional responses to the death of things we love.

I am not rejecting the hospice model as Woodbury hoped it would be understood. But I wonder whether most of us are just not equipped to imagine a hospice – most of us have not been inside one, and an image is, after all, only as useful as the meaning it conveys. We imagine hospices as places of death, not healing, and certainly not hope. The image is powerful because it resonates with our deepest fears, gives us something to hang them on, and makes us feel less alone. Hospice care is undeniably one part of the work we are called to. Defending and dignifying the dying is an act of faithfulness and compassion, and there are many places and creatures that it is already too late to save. The vocation of palliative care humanizes our communities.

Yet the world is not homogenous. Not everything is dying, and not everything can be cured in the same way. This cannot be the only way we imagine ourselves – a reading of the hospice model which rejects the power of acts of healing is not courageous, it is cowardly. It assumes that our worst nightmares are the most real. And to imply that the palliative work of hospice care is somehow opposed to seeking a cure makes the nurse an ally of the disease. Woodbury proposes hospice care as an acknowledgement of 'the end of life as we have come to know it'. We do not ignore the range of frightening outcomes we face. But it does mean "avoiding unnecessary fatalism regarding those future outcomes, and dealing instead with witnessing and accommodating what is unfolding in the present moment."

I am interested in how Christians should witness to and accommodate this present moment. Let us imagine then that we are not in a hospice; we are in a hospital. It is a familiar hospital: it is stretched beyond capacity, with trolleys up and down the corridors, sleep-deprived nurses, and high risk of infection. Yet it is still a hospital. Hospitals are still sites where the terminal are comforted. But they are also sites where life is snatched back, where the dying return to us, where the most vulnerable fight to survive. How dare we give up on them before they have given up on themselves. How dare we! Hospitals are places where battles are lost, yes – but the war is never abandoned. Nihilism is not an option. As Heglar puts it:

It is absolutely possible to prepare for the disasters already, terrifyingly, upon us while also doing our damnedest to quit baking more in. We can acknowledge the storm of emotions that comes with watching our world unravel, process those emotions, and pick ourselves up to protect what we can... We don't have to be pollyannish, or fatalistic. We can just be human. We can be messy, imperfect, contradictory, broken. We can recognize that 'hopelessness' does not mean 'helplessness.'

Palliative care is undeniably a Christian vocation: dying things still belong to God. Every time we treat our fellow creatures – and the rest of the living world – with kindness, even in the face of death, every time we relieve the suffering of even a small part of creation – we give dignity to the dying. God is indeed calling His people, now, at this period in history, to a ministry of hospice care as part of our witness to the world. Yet a hospital model invites us to humility, not hubris. We may hold a cure, but we are not responsible for diagnosing every ward. A nutritionist might be able to propose that a certain diet would be helpful *generally* for many different kinds of health problems, but they should not go around suggesting that eating vegetables can cure cancer. I wonder whether the Apostle Paul would remind us that a body made up of eyes is a useless body. Our climate activism must, first and foremost, let go of the desire to insist that everyone responds as we do in order to participate in meaningful work. That way despair lies.

A hospital is not just a place for sick and the dying. It is also a place where a new world is born. In the hospital, the end of life and the beginning of life are bedfellows. Our model of hospital care will find real meaning in a belief in God's redemptive power. The subtitle of Woodbury's hospice paper was 'Rebirthing Planet Earth', a point almost entirely ignored in the headlines that followed its release. Woodbury is a Buddhist, not a Christian. His understanding of 'rebirth' is very different to mine, but we share a conviction that death is not the same as destruction, and indeed that death often makes way for renewal.

I cling to the resurrection in light of, not in spite of, my feelings of despair. The world was sick and needed healing 2,000 years ago. Jesus did not instruct his disciples to save the world or condemn it to death. Instead, we take up the call to compassion, healing, dignifying, and defending those we encounter in anticipation of what God has already done and will do. To put it another way: our context is unique, but the work of the Church here and now is not. We have already been called to heal the sick, care for the dying, and pursue life in all its fullness, knowing that Jesus has already made things well again. We are both palliative nurse and midwife. Resurrection is coming.

Hannah is an ordinand in the Church of England and a PhD student on theologies of grieving nature. She set up the Christian Climate Action group in Manchester.

28

Contemplating extinction

JEREMY H KIDWELL

I've often found that when grassroots work is most successful, the experience can be electrifying. Activism is all about mobilising people around common or overlapping concerns and big ideas. Yet it is important to remember that it can also be about a meeting of hearts, as we approach issues on the basis of a common sentiment: whether hope or terror. There is a dark side to affirming the felt side of activism inasmuch as issues can become strangely fuzzy. This is particularly the case with extinction. As I have attended climate marches and now Extinction Rebellions across the country as a scholar/activist over the past decade, it has often struck me how a range of divergent concerns can coalesce under a common banner like 'environmental crisis' or 'extinction'.

What exactly do we mean when we say 'extinction'? When I've asked this question of Christians and other faith-based activists, the answers have ranged widely. Some people have in mind the extinction of specific local plants, animals, insects, or birds; for others it is the eradication of whole species. Still others are thinking of the disappearance of indigenous populations worldwide; or at the most severe, the disappearance of the human species from Earth. On the one hand, we can appreciate that these extinction concerns are intertwined. But on the other hand, I want to ask whether these different concerns carry us into action in different ways. There are different motivations or emphases lurking here: is our concern local or distant? Are we talking about the extinction of our own species or other creatures? Our work under this big tent and our active pursuit of solidarity can conceal these kinds of differences. As I'd like to suggest here, these different interpretations of extinction are important inasmuch as they carry us into different kinds of spiritual practices.

By contrast, when I ask people to describe how they *feel* about extinction, there is a much more consistent response. We feel deeply sad about the state of our earthly home, often to such a level that this work functions as a kind of lament. To a great extent, the work of environmental activism has in past years been procedural: about mobilization, campaigning, and trying to promote a response to scientific observation and communication. Only recently has the global community begun to appreciate the importance of the way that things like extinction *feel*, and recognized the importance of those communities which nurture reflection on the felt experience of the crisis, including churches. As many readers will relate, we can feel paralysed, out of place, or hollowed out from the regular confrontation with environmental issues. So I want to underline the importance of more holistic activism which attends to emotional work at the outset, before digging into some of the complexities which lie in the midst of our felt response to these extinction crises.

To sharpen my question a bit further, what is it that we are *mourning* with the issue of extinction? Is it the loss of specific species? Like Martha, the last passenger pigeon? Few of us will have had the privilege to live in biodiversity hotspots to tangibly and relationally witness species decline, so it is unlikely that we have ever had an actual embodied experience of these extinctions. Though our mourning may be sharp, the object of this outpouring of emotion remains fuzzy. Here, it is quite possible that lament might serve as a sort of hiding place or proxy for our lament of other deeply felt personal losses such as the loss of 'home' or the loss of access to rural landscapes. Given the way that so many of us are now urban and highly mobile, our primal loss might be something much more abstract, such as the loss of a feeling of familiarity towards our local world as climate change makes the experience of different seasons or places feel suddenly and sharply foreign. Our daily lives are now full of much more frequent changes of home, career, relationships, and community in a condition that Zygmunt Bauman has called 'liquid modernity'. Is it possible that on some level we are lamenting the absence of stability and stasis itself?

This last subject of lament becomes even more interesting when we consider how, even more paradoxically, we may be mourning things which are not yet lost. The trouble with mass extinction is that as an event, we can't be certain it has happened until decades or centuries afterwards. So it is possible that the event which Elizabeth Kolbert has described as the "sixth extinction" has not yet occurred. Instead, it remains a very concerning projection, just on the brink of happening. On a more individual level, many of the creatures which we lament are not *gone* but are *nearly* extinct, like the so-called 'ghost species' who are on an unavoidable trajectory to extinction as their habitats have already been permanently reduced to a size or condition which cannot sustain future generations. Theologians use the word 'proleptic' to speak of this kind of anticipatory concern. I'd be the first to admit that much of my own lament is caught up in proleptic elegies. In my experience, this is a kind of mourning which is not diminished, but actually intensified by the fact that I can watch helplessly as these creatures and habitats proceed on a seemingly inevitable path towards destruction.

There are also some hazards lurking here. Particularly in the ways that our processing of eco-grief can be caught up in the kinds of privilege that people like me enjoy. The historian Patrick Brantlinger, in his study of the British colonial project also uses this term "proleptic elegy" to describe forms of mourning for far away things which might yet be lost. However, he points to the way that recurrent proleptic mourning for things far away might also serve as a kind of liturgy which trains people to mobilize paternalistically in response to these imagined threats. If we believe that a creature, community, culture, or place is about to disappear forever, more radical interventions seem far more appealing and may also receive less scrutiny. The result of this was the generation of support for the unnecessary and harmful take-over of whole communities in far off places, particularly in places like Tasmania. Personal grief can be both a strong and persistent source of motivation, and these past events remind us that personal grief for an imagined loss can underwrite processes we wouldn't ordinarily support, such as the wholesale disenfranchizement of other persons.

Are there sharp, perhaps hidden edges to our campaigns which arise from this kind of fear and are not adequately chastened by compassion?

This sensibility can also implicitly underwrite exclusion. As other authors in this book have highlighted, the impacts of extinction, climate change and habitat loss are highly variable, with impacts falling more acutely on people in the global South and within specific places, upon those with fewer financial resources. So while I might experience climate change as a sudden and novel disruption of everyday comfort and security, the relatively recent occurrence of this sensibility is the product of white privilege. In my experience, I have realized in retrospect that some of my eco-lament can serve as a more noble stand-in for white fragility. Kathryn Yusoff writes of "A Billion Black Anthropocenes or None" in order to highlight the problems of exclusion which have been present in environmentalism.

There are two consequences of this which may be relevant to other white campaigners and should be highlighted here. The first lies in the tendency, especially for white Christians, to assume implicitly that our own challenging experiences are universal. Whilst climate change is a global problem, its impacts are unevenly distributed. To put this another way, in our lament, we may miss the fact that our suffering could actually be much worse, and that others are already contending with far more serious impacts. So whilst I may be (rightfully, I would emphasize) lamenting the inevitable near-future loss of beautiful landscapes which have been important to me, others are lamenting the active disappearance of their homes, livelihoods, and subsistence. I'd like to suggest that, now that we have finally sharpened the sense of threat and emergency in the public discourse, white Christians like me need to spend some time reconceptualizing solidarity.

One way to do this is to set aside our sense of universality of experience and listen to non-white activists more attentively. This is particularly important, not just because it will help to add a level of badly needed intersectionality to our activities, but also because those persons are likely way ahead of us in developing forms of resilience and response to extinction. A second practical outworking of this concern as a form of attentive listening is that white activists need to think more carefully about who they speak for. Solidarity is a crucial aspect to effective mobilization, so there is often a desire to speak for 'the earth' or the 'global community'. White justice-oriented Christians are used to speaking up for victims of injustice, and this is a good impulse, well justified as a Christian discipline. However, there are opportunities to let those victims speak for themselves. So whilst we attend to our own grief as an important aspect of the work of contemplating extinction, let us be careful to note where these losses are real and where they are forecasted, and be careful to reckon with the way that loss is tangled up with our personal experiences and entitlements.

I've watched with interest as a number of recent protests, among them Extinction Rebellion, have been a laboratory for revealing the lack of intersectionality in eco-protest. In particular, people of colour have highlighted the ways that casual, even enthusiastic, engagement with police in the midst of civil disobedience invokes white privilege.

Exposing these fault lines is a good thing and long overdue. It's the opposite of failure – rather the indication that a very challenging process of maturing our activism and our collaborations is now underway. If anything, this suggests that we may finally be arriving at a moment when white Christians have an opportunity to do the work of reconceptualising solidarity. Are we speaking for someone else when they have a voice that might be listened to and amplified instead? Or, in a similar vein, is our mourning as comprehensive as we *feel* it to be?

To return to my broader concern here, around contemplating extinction, I want to suggest that there is some important work to be done here, not just at the picket line, but in a much wider sense of cultivating the forms of spiritual discipline which might underpin compassionate and inclusive earth-focused activism. I think there hasn't been an adequate appreciation of the ways in which many environmental activists and campaigners are engaged in ministry as "wounded healers", as Henri Nouwen put it. So here we might try to emphasize more fully the important role of self-work as preceding and underpinning our more performative occasional actions, just like any minister or chaplain practices spiritual discipline throughout the week in order to preach and minister.

Sometimes the (very welcome) emphasis on the urgency of action to prevent extinction can make activities like contemplation seem frivolous and unimportant. What I am trying to suggest here is that there is some important work here to be done linking up the facilitation and leadership we'd like to bring to our communities on the issues of climate change and extinction with some of the self-work that is required to develop personal resilience as we grapple with the shocking scenarios that are represented in both extinction and a climate emergency. One of the XR core values which activists often highlight, but which doesn't appear as much in the media is the work of fostering 'regenerative culture'. This kind of work is not a matter of veering away from the urgency of these crises, but rather about recognizing the need to weave ongoing practice of the disciplines of silence, solitude, and meditation into this vital work.

Jeremy H. Kidwell is a Senior Lecturer in Christian Ethics at the University of Birmingham.

29

How a Franciscan Brother found XR

BROTHER FINNIAN

My name is Brother Finnian and I recently participated in two Extinction Rebellion protests in London. I got involved because I am increasingly concerned about the climate crisis and its impact on the poorest people in the world. I don't consider myself to be a very political person, but the climate emergency connected with my underlying motivation to serve the rejected and marginalized, which I now understand to include creation.

I am an Anglican Franciscan brother living in Plaistow, in east London. I live in a community of five brothers and nine homeless adults. We give out food at our front door to at least 50 people each weekday. Four years ago we gave out food to two or three people a week, revealing how quickly food poverty has increased in the area. My community also runs drop-ins from 9am–5pm each weekday, primarily for people who are socially isolated or experiencing difficulties securing accommodation. My life revolves around a daily cycle of Christian prayer, and offering hospitality to the stranger.

I first became aware of a specifically 'Christian' concern for Creation when I was a volunteer at Hilfield Friary. While living there I learnt St Francis of Assisi had composed a prayer called 'The Canticle of the Creatures', in which he presented creation as praising God. I hadn't really thought a lot about this until a year ago when I attended a week long Formation on Pope Francis' encyclical *Laudate Si*, at the Roman Catholic Poor Clare convent in Arkley.

It immediately struck me that this encyclical was directed to "all people of good will", rather than to only Roman Catholics, or even to other Christians. Pope Francis was addressing everyone who had a concern about the climate catastrophe we are now facing. Pope Francis begins *Laudate Si* with a short reflection on St Francis' Canticle of the Creatures. During the Formation we were told St Francis and the Franciscan Movement was made distinct by emphasising the filial relationships between all created things. Franciscans have always understood that the Lord granted humanity 'stewardship' for creation rather than 'dominion' over it (Gen 1.28). We see this in the Canticle of the Creatures where Francis uses the titles, 'Brother Sun' and 'Sister Moon'. Franciscans have consistently seen themselves, and humanity as a whole, as being located within creation rather than standing over it.

Today the principles of Saint Francis were read after our Morning Prayer. We were told brothers and sisters, "will rejoice in God's world and all its beauty and living creatures". I suddenly realized my underlying motivation and desire for being part of the environmental movement is to reveal the beauty of God to people. I think it is only when we know God, and God's Creation, as being 'beautiful' that we will really address the climate emergency.

Franciscan pastoral work has often emphasized the

inherent 'beauty' and value of the person we seek to minister to, whoever it may be. We see this in St Francis' work with the lepers, and it continues to inform how Franciscans engage in a variety of issues ranging from chaplaincy in prisons and hospitals to giving out food and clothes to people experiencing homelessness. Just as we house our homeless brothers and sisters in my friary, and feed our hungry brothers and sisters at our front door, we also seek to serve our brothers and sisters within creation who are suffering at this time.

I attended the Extinction Rebellion protests with two other members of The Young Franciscan Community, a new lay Franciscan community in West London. This new community has three adults in employment who share life together, and pray several times a day together using the SSF Office book. There are also 13 live-out members who attend regular prayer meetings, and evenings of Formation in Franciscan spirituality and practice. While there we also met members of other Franciscan groups.

I've loved meeting other Franciscans getting involved with Extinction Rebellion who are also feeling called to non-violent activism. If you are within the Franciscan family, or are someone who looks to St Francis (and Clare!) for inspiration, I would encourage you to think about joining us too. Speaking out about injustice is part of our charism and these peaceful events are a good witness to non-violent forms of political activism.

Pax et bonum!

Brother Finnian is a member of the Society of Saint Francis, an Anglican religious community.

'Love and ever more love is the only solution to every problem that comes up.

If we love each other enough, we will bear with each other's faults and burdens.

If we love enough, we are going to light that fire in the hearts of others.

And it is love that will burn out the sins and hatreds that sadden us.

It is love that will make us want to do great things for each other.

No sacrifice and no suffering will then seem too much.'

Dorothy Day

30

In whose hands?

SAM WAKELING

Climate breakdown is an illusion.

Not because it isn't happening. But because it is.

Because it creeps in with a glacial slowness, spread across a vast distance, and in countless tiny incremental ways. Sea rising, drought lengthening, diseases spreading, insects disappearing. Each isolated effect, each warm winter day, could seem coincidental.

All of these make it hard for us to even perceive.

Harder to absorb.

And hardest still to act to resist in a way that matches the situation we are in.

We look around each day and things seem normal. The sun rises. The rain falls. Good and bad things happen. Trivial irritations and distractions are always available to fill our attention. Life just goes on – or so it seems.

Yet like a throbbing, ringing in our ears, we know otherwise.

With bleary eyes we try to fight ourselves awake. When the long, slow, thinly-spread, bit-by-bit horror of what is happening to our living world starts to come to life. When the decades of recent history and the ocean of years ahead concertina down into this awesomely pivotal present. When the thin veneer of normality starts cracking.

When keeping to 1.5°C of warming is now only a handful more years of current emissions. And those emissions are still going up, not yet even beginning to drop.

So our attention returns to the closest parts of us. We're reminded of our tiny, fleshy part in this wide horizon stretch of living things. Our molecules and cells teeming together in every breath, every heartbeat. And laid in front of us we find the ever-present tools we've been given.

Our hands.

You can hold them both in front of you. Make one into a tight fist, clenched. Keep the other as an open palm.

Think about this contrast. Dwell with it for a moment.

What feelings do they each represent?

For me the fist is: fear, anger, violence. Keeping hold of what we have, holding onto our power.

While the palm shows welcome, generosity, connection. It is vulnerable, fragile, and undefended.

They show two different ways in which we could react to the unfolding crisis.

I have often responded with the fist. White knuckled, fingernails digging in, quivering, unstable.

Fear, at what a collapsing, destabilized world may bring for my children. Anger, fury, at how we've been brought to this, most of all the poorest, least responsible, by people in power over the years. Men with wealth and control, making decisions with the knowledge of what is happening, but choosing to ignore the consequences. Still hoping to greenwash and deny rather than act to defend life. I find myself wishing I were able to control things, but feeling powerless.

I think I've known climate change is a serious thing for years. Like many, I've tried to do what I can, hoping that would help. Living simply, as humbly as I can. But somehow, in the last year, the sadness and loss of it has come home. I've felt broken, hollow. I've found myself crying uncontrollably. Privately and sometimes in public.

At night, holding my sleeping baby, I'd wrap my arms tight around him. I'd try to promise him and his sister that I would do whatever I could to protect them. Even though I know that can never be enough.

And it's on my knees in this Gethsemane where I'm met again.

By a gentle open palm of freedom.

A place to hold these fists. To show them without shame, to know that they are understood. To hear that these wounds we carry aren't hidden, ours alone – but shared by the One who knew them first. More than we ever could, the Creator knows the pain and loss of witnessing the destruction of that which they had given life.

And hearing that still, small voice, there's a whisper which begins to release my fingers. To coax and caress, to open slowly into a blossoming palm.

"You'll learn to love as I have loved you."

To find less and less that I want to keep hold of, and more readiness for everything to change.

To meet violence with vulnerability. To lay down our lives. To love our enemies.

To meet a rising tide of fearful fists, of a world which knows only how to hold on tight, with a defiant, fragile welcome.

And so we face a system of domination, government and economy — being both led and followed by these clinging, timid fists. Afraid of losing what we have, and scared of change.

And so, in denial, we talk of plastic straws as we build bigger airports.

In October we took our children to London again. To walk the reclaimed streets where grinding, monotonous, poison machines were replaced with song and dance and courage and sacrifice.

For my wife Louise to sit in the road in Whitehall, with many other mothers. To give the most natural of care: to feed our baby his milk.

And for me to join a sit-down vigil of prayer and worship at the entrance of London City Airport, where government investment is aiming to double flights by 2035. We sang of amazing grace, and of sorrow and love flowing mingled down. We prayed "Your will be done on earth as in heaven", and sat together in silence.

And we were dragged away by the Met Police.

While we are trying to learn the way of Jesus, of how to hold our palms open, we find what he knew all too well: that our authorities still find the fist of control and fear to be their only tool. They can only try to keep hold of their power.

But what this movement says – following many others

using nonviolent civil disobedience, like Martin Luther King – is this: whatever force we are met with, we can choose peace. We can use our fragile bodies and our discipline. The divine fruit of patience, kindness, goodness, faithfulness and self control ("against these there is no law") can expose and undermine the power of that fist in some mysterious way.

As we were led from the police van to the holding cell, two of those with us, who were over 80, were shuffling slowly. One young officer turned and stifled a bewildered laugh, "I can't believe we're bringing in you lot."

And another crack opens up. As Leonard Cohen sang, "That's how the light gets in."

We can offer our weakness to God and say: we're here, we're broken, but we're open. And we want to do this. We want to come together. To leave behind whatever it takes. To walk with You to find justice and mercy, to protect and restore life.

Each day we can remember our hands. We can reject the illusion that things are all fine.

We can continue struggling to come more fully awake to the wonder of this world, and the echoing depths of loss being carved into it. We can recognize when this causes us to ball up our fists, and meet God there. Listen for the quiet invitation to unravel, to be transformed into His new creation. Take another step into resurrection.

We are invited to follow, holding His hand. A hand carrying the deepest of scars, but which He still holds open for us.

Sam Wakeling, father of two and civil servant living in Sheffield.

Richard Barnard, locked on to fellow CCA protestors, awaits arrest on Blackfriars Bridge.

THE
HANDS

31

How can the Church support its activists?

CHRISTIAN CLIMATE ACTION

It can be lonely being an activist in church. We might get an occasional slot in a service, or a space on a notice board. But whether our particular concern is fair trade, or people trafficking, or the climate, the issues that fire us up can often feel marginal to the church. Sometimes they feel entirely unwelcome, a distraction from the more traditional priorities of worship and teaching.

That gets even more difficult when non-violent direct action is involved. Once church members are breaking the law, it gets controversial. Not everyone will agree with the approach. Christians aren't supposed to get arrested, are they? We might get labelled as troublemakers. We might sense a distance between us and people we previously considered friends. Some might see a tension between our activism and our other work in the Church: should this person be trusted with the young people? Can they still be an elder? Perhaps they should be quietly dropped from the worship rota. In some cases, members of Christian Climate Action have even faced church discipline.

As activists, we feel a call to action that we cannot ignore. Like Jeremiah, there is a fire in our bones. Many would prefer not to break the law, but nothing else has worked. This is what it has come to, and we are well out of our comfort zone. It can be stressful and burnout is common. If activists find themselves under a cloud at church too, that stress is compounded.

Activists need support, encouragement and care. The church may or may not be able to find consensus for formally supporting Extinction Rebellion – that's another chapter – but it can certainly support its members that are involved.

To investigate how the church can support its activist members, we asked them, in and around the Christian Climate Action tent in Trafalgar Square during the October Rebellion.

Talk about climate action

"Being open to dialogue is a good place for churches to start," says Hannah Kittle from CCA Leeds. Talking openly about activism avoids rumours and misunderstandings. "Find out about people's reasons and motivations for action. Otherwise it's easy to write people off as extreme in their views."

Rachel Webbly, a team rector from Whitstable, knows what it's like to have activists in her church. "There are people who have been quietly supporting Greenpeace, one lady who's done that all her life. We talk about it regularly." Involving the whole church in Extinction Rebellion would be seen as too political with her congregation, so "it's really about encouraging people. People have a lot of things going on with their lives already, and it's counter-productive to burden them with guilt or fear. That's not the point of church, but we do talk about it and pray about it. For example I've brought a few prayer flags with me that people made, and they will know that they have contributed to the prayers today."

Encourage action

Talking isn't enough on its own though, says Hannah. "We need to see the value in acting as well as speaking. I feel like there's been a shift recently in the honesty of prayers for the climate, not just thanking God for the world, but acknowledging that we have pillaged it and done serious damage. Seeking forgiveness for that spurs me into action. I want to do something as well as asking forgiveness. Real repentance is about changing, turning and acting in a different way."

There's a difference between recognising the calling on some people's lives, and actively encouraging them to pursue it. That matters in climate action, just as it would if someone was called into overseas missionary work, or full time ministry. "Climate action isn't just legitimate," says Hannah. "It's important. It shouldn't just be tolerated. It should be encouraged."

Lead from the front

Encouraging climate action publicly as a church is a matter of leadership. That could of course be a minister who is an active member of CCA, but it could be through sermons, prayer in services, commissioning activists ahead of major actions, etc. When climate action is seen as a fringe concern for those who are 'into that sort of thing', a little leadership from the front can go a long way. "There's a real authority and authenticity that comes from having staff or clergy on board" as one CCA member said.

Little gestures don't go unnoticed. Another activist noticed that "the church pastor 'liked' my Facebook post saying I was coming here, so I guess the church is on board!"

Pray for your activists

"It's important that people should talk to each other and listen to each other, and as a person of faith I think it's important that they should pray for each other," says Paul Bayes, Bishop of Liverpool. "If you're someone for whom this is really crucial, and something you've given your life to, then it would great if you can share that with your local congregation and ask them to pray for you. And more than anything, if you're praying for someone who's out on the streets taking part in actions like this, and afterwards you talk to them about it, that will raise your consciousness too. So mutual prayer and support is a great thing."

Asking someone to pray for you gives them a role too, a way of being involved in your action and standing with you. Not everyone is in a position to travel or take part in climate action, but as Rev Jo Richards from Canterbury pointed out, "everyone can pray."

Provide pastoral support

Being involved with civil disobedience is hard work, physically, emotionally and spiritually. It is really important that activists are able to talk, pray, reflect on and process their experiences, something the ministry team at church are well placed to help with. One approach could be to provide a space for activists in your church to debrief and pray together. On a simpler level, church members could have activists over to dinner and ask how things are going. Look out for them the week after a big action; they can be powerful moments, and coming back to 'normal' church can feel deflating by comparison.

"When I got home I started feeling the effects of this action and the arrest," writes one CCA member. "I felt vulnerable and for some days regularly cried uncontrollably without really knowing why." It can take time to process the intense experiences of civil disobedience, and activists may appreciate a listening ear and somebody to pray with. They may need a break from normal church responsibilities and a bit of space. Court cases might not get as much attention as protest actions, and can be a time when activists feel more isolated. Standing with people at this stressful time is particularly important.

Recognize the role of activism in growth and discipleship

As people develop awareness of ecological crisis and begin to take an interest in activism, the church can help them on that journey. "As a minister we're often midwives of other people's growth," as Rev Mark Coleman puts it. That is something that can be nurtured through teaching, or by putting people in touch with like-minded Christians. "We can connect people in our congregations with Christian Climate Action, Green Christian, or others in those networks, and let that flourish. Then those people can come back and enrich the worshipping body, telling stories and showing pictures of what they've been doing."

Recognize climate action as mission

Christians have played key roles in Extinction Rebellion, with CCA members involved in some high profile actions. That is a real demonstration of Christian sacrifice, and love for others. There are often opportunities for activists to talk openly about their faith and what motivates them. Clergy in collars or robes are a visible Christian presence that is noticed and appreciated, and the welcome to prayer, worship and communion can be powerful too.

"When we're worshipping in this context, we're evangelizing," says CCA member Samantha Lindo. "Here is where we're testifying. We hear people going 'yes, the Christians! The Christians are here!' It's so relevant – out here we are the minority, being the salt and the light. This is where we're called to be."

Bring the teaching

Many of the hesitations around direct action or even around the environment come from gaps in church teaching. Mark Coleman has been preaching about these issues for a long time, but recognizes that he has been on a theological journey. "I'm learning slowly to notice things in the

Scriptures that I hadn't seen before – all those references to nature that I'd somehow tuned out before. It's just been part of the wallpaper, whereas now I see and relish the natural world that Jesus operated in. There are theological journeys I've been on, and that the church as a whole needs to take."

Samantha agrees that "we need to get the theology clear. Climate action is part of what it means to not be lukewarm, but to follow Jesus at this pivotal moment in history. It's not business as usual any more, and we can't twiddle along in our individualistic worlds and think that it's what life is. The best thing is to preach that theology, and start collectively challenging our world and our values."

32

How my church supported the rebellion

CANON GILES GODDARD

What did St John's do during the rebellions?

In April we opened the whole church. It was during Holy Week, so the building was only being used for services. Usually there are lots of other things going on, but as it was quite quiet we were able to open up the church and the crypt for Extinction Rebellion activists to come and stay. Basically we just made them as welcome as we could.

Rebels used it as a campsite, and a space for rest and recuperation. We're very close to Waterloo Bridge, so people were able to come and go from the Bridge to use the toilets. We have a kitchenette that people could use downstairs. We could offer internet and electricity. We're lucky that we've got showers here, so people were able to use them and really appreciated that. Not every church has those.

We did the same in October, but because the church was more in use we just opened up the crypt. We still had quite a lot of people sleeping there, and they were very careful about looking after the place, keeping it clean and so on. At some points we had 60 or 70 people sleeping here, so it wasn't easy – but definitely better than sleeping out on the streets. We specifically hosted the Red Rebels – that's the performance group with the red costumes and white faces. They made their way from site to site, and they were based with us.

How did it go?

The congregation were very supportive, partly because lots of them aren't on site in the week and it didn't affect them, and the staff were supportive too. The people who found it harder were the regular occupants of the church, and the organizations who rent office space from us downstairs in the crypt. In April they weren't around so much and it worked fine. It was more complicated in October because the offices were all being used as well, and there was quite a lot of pressure on the space. Some of the regular users found it a bit difficult. And it was for two weeks. They were okay, but by the end of the second week tempers were beginning to fray – two weeks is a long time!

I did get some complaints from outside the church. We flew the XR flag, and I had one person who was very cross about that. He'd had a difficult commute, and rang me up to shout at me! I realize that we can be in a bit of an XR bubble here, and not everyone sees it the same way. Overall though, it's amazing the level of at least tacit consent there is for what Extinction Rebellion are doing.

We had some extraordinary responses from the people we hosted. Because of the XR approach, which is very much about working alongside people, they were so grateful it was

I've been completely blown away by Church support during the two-weeks of action in October.'
Olly, Extinction Rebellion

'I'd love it if all the churches where these actions happen could open their doors and give people shelter. We are in many ways pilgrims, and giving hospitality to pilgrims is something churches have done for centuries.'
Robin, Christian Climate Action, Cambridge

'St John's has become our sanctuary. We feel safe here, and looked after. And we're all incredibly grateful.'
Ruth, RedRebel, Stroud

'I'm a Buddhist, but St John's has reaffirmed my faith in our ability to cooperate on a human level, set aside differences and work together. I can feel the inclusiveness and passion in this place.'
Luke, Extinction Rebellion Brighton

almost embarrassing really. I kept having to say "No, it's us who should be thanking you". I had a very nice letter afterwards from someone saying how touched they were that she and her daughter had been able to sleep here. We had a lot of conversations where people were saying this is what the church ought to be doing, or with people who had never experienced church like this before. We've even had a couple of people join the church as a result.

Were you able to get involved on the streets as well?

I didn't do as many actions as I would have liked to, but six or seven of us from St John's helped to block the bridge on Monday morning. We were on the Faith Bridge, and then I visited various other sites during the week. But my first responsibility was to be here at the church, and making sure it all worked here. It felt like the sites were well covered and had enough people, so I tried to be where I was most useful.

What advice would you give to other churches hosting rebels?

If we were doing it again we'd certainly need to put some clearer boundaries in place, and do some advance work with the groups. Engage early with other users of the building. Try and set some limits in terms of time and space. And frame it very much in terms of mission – this is how we can support this urgent work against the climate crisis.

What was really helpful was that someone from XR, Andrew, served as the point person for the site, and they appointed two guardians, Luke and Liz. They were the people who liaised with me, and with the staff team who look after the church. I didn't always remember to report back to my staff what I had agreed with XR, but that was my internal communication problem! It definitely helped to have a single point of contact. We didn't have that in May, and it was a good learning point for October.

There's nothing in canon law that stops us welcoming people. Our insurers were fairly relaxed about it, so the main thing is having clear lines of communication. Our guests were generous too. We had quite a substantial donation at the end to help cover the cost of electricity and other things.

I don't know what the future holds for XR, and whether they'll be doing the same sorts of actions in future. But I definitely do know that the payback in relationships, and in how we are perceived, well outweighs the inconvenience.

Giles Goddard is the vicar of St John's Waterloo, one of several London churches that hosted activists during the Spring and Autumn rebellions in 2019.

33

Arrestable?
Three stories to help you decide

JO RAND, MARK COLEMAN
AND HILARY BOND

Arrests are a central part of Extinction Rebellion strategy. Being willing to lay down our freedom for our beliefs is a powerful statement. As Christians, we are called to sacrifice, to offer ourselves for others. As we risk our liberty, we show that we are serious. We sound the alarm, and we bear witness against injustice.

Most activists in Extinction Rebellion will at some point consider the question – are they going to risk arrest as part of their non-violent direct action?

For some people it's not really on the cards. That may be because of caring responsibilities, citizenship status, or medical needs. We all have different vocations and different circumstances. We are all at different points in our lives and our careers. Some face greater risks, while others are in a more privileged position. The emotional toll is different for everybody too. Some find it distressing and disorienting, and some feel a sense of peace about their actions. It's not possible to predict how we will respond in advance.

With all of that in mind, risking arrest is a matter of wisdom and discernment. Nobody in the movement should ever feel pressured into an arrestable role. For those in a position to take that step, it can be a powerful statement. We invite Christian activists to seek God's wisdom on the matter, with courage and prayerful consideration.

Here are three stories of people who were arrested during the Faith Bridge actions, as part of the October Rebellion. They are all ordained ministers, but many of our members who get arrested aren't and it's certainly not a prerequisite!

Jo Rand, a Methodist minister from Cumbria, was arrested for her role in taking and holding the Faith Bridge at the October Rebellion.

I had come down thinking that as the week went on, I was prepared to get arrested if that seemed appropriate. I wasn't intending to be arrested right at the beginning, but that's where it ended up.

The police asked me repeatedly to move. I said no. I decided I wasn't going to cooperate and get up and walk off the bridge, so they did that thing where four police officers take a limb each and I was just going floppy in the middle. I felt a huge sense of peace about what I was doing, about what was happening to me. That surprised me. The process of being on the bridge, of being arrested. I felt completely calm, that I was in the right place, and where God wanted me to be. I've had a real sense that God is calling me to be part of this.

Jo Rand with CCA in Trafalgar Square

It took quite a long time for transport to arrive, as they were clearly busy. They took me and one other from a different group off to the police station. My details were taken, and I was taken down to a cell. I asked for a Bible, and so it was me, by myself – time to read the Bible, time to pray.

It was strange feeling to sit in that little room and know that I couldn't leave. Sometimes I could hear other people in the cells who were feeling more agitated, and I just thought "I'm here. God is with me. I'm alright."

I was there quite a long time. They take your watch off you, so I don't know what time things happened. It was about quarter past five in the morning when someone came to get me and said I was being released, pending investigation.

One of the stories from the Bible that has spoken to me is where Nathan challenges King David over his actions in taking the wife of Uriah. Nathan has to confront him, and he tells him a story about a man who has flocks and herds and so many things, and yet he sees this poor man who has one lamb, and he has to have that lamb. Nathan challenges David and says, "you're that man". It strikes me that this is like what we do in the developed world. We have all we need and more, and yet we look at the resources of the planet that are in other parts of the world and we say, "I want that".

My congregations have mostly been very supportive, but I am finding it is having an impact on being allowed into schools – one has forbidden me from mentioning my involvement in the protest, even though the whole school knows about it. At the time of writing another has not yet let me return to take assemblies, pending a conversation with me at the next governors' meeting. I do not yet know whether I will be interviewed by the police or charged, but I don't regret what I have done, even if it restricts what I can do in future. This is an issue of justice, and my faith compels me to act.

Rev Mark Coleman travelled from Rochdale to take part in the October Rebellion.

I have been involved in environmental campaigns for over ten years as a Christian priest. Like many, I have signed petitions, I have gone to events in London, I've been to see my MPs.

Over that time, the emissions of carbon into the atmosphere have increased. Many words have been spoken by our national leaders, but not enough has changed. We are seeing the impacts of one degree of warming, and are now in the midst of a climate emergency and the sixth mass extinction of species.

I was arrested for the first time in my life on Lambeth Bridge.

I was nervous before I was arrested, and then I saw a fellow clergy person and I sat next to her. Another man said "You'd do anything to get out of another church meeting!" and we laughed about that. Then I asked if he'd pray with us, and we did.

Together we got arrested, and it was a profound experience. There were some other lovely activists who have been doing this for years, anti-fracking and so on, very special people with a deep love for the earth. The police were very relaxed, and it really felt like we were in a sacred space in that police holding room as we sat together on the hard floor, sharing biscuits and stories.

I do think we who represent the church with a collar need to get out there and be seen – though this is for everyone, clergy or not.

Don't be afraid of being arrested. We're privileged people when we're in ministry.

I hope and pray that my arrest, along with that of over 270 others that day alone, may help make the climate emergency a priority for politicians, so that my parishioners in Rochdale may have a chance of life in all its fullness.

Reverend Hilary Bond, from Dorset, initially took part in a 'non-arrestable' capacity. She explains how that changed on the day.

This was what happened on the afternoon of 7th October 2019. People were taking turns at the microphone and speaking or praying or encouraging. Then a large group of police appeared in the middle of Millbank. We decided to sit down, and the police began to move towards us. I sat and watched as a group of police moved through the crowd, dismantling and carrying away all the kitchen equipment, including the food. Many people were moving as we realized that the police were going to try to move us back and that arrests would happen. Then the police began to approach people one by one, asking them to move.

At that point, CCA member Helen (also clergy and who I got to know later) came and sat next to me and made a comment about arrestable clergy. She looked as nervous and unsure as I felt at that moment. A young police officer came and squatted down in front of me. He asked me something like "Why are you doing this?" That was really helpful, because I knew exactly why I was doing it.

I said to him "I am doing this because it matters. Because we only have 12 years in which to try to put right the devastation of climate change. I don't want in 20 years' time for it all to have gone horribly wrong, and my children and maybe grandchildren to come to me and say 'you knew! Why didn't you do something?' At least I will be able to say to them that I did what I could."

He briefly suggested that there were better ways of doing this than being a nuisance and inconveniencing so many other people. Then he said, "I am going to ask you once more to move over to the side. If you don't, I will have to arrest you. Do you understand?" I said that I did understand.

He asked me again "Will you please move?"

I knew in that moment that if I moved it would be as if I was giving a message to the whole world (if it cared to be watching!) that I cared about myself more than the rest of God's creation – and it's not true. So I said, "I'm sorry, I can't", and he arrested me.

Hours later I found myself kneeling on the floor of a police cell overflowing with thankfulness and peace;

powerfully aware of God's presence and the knowledge that I was exactly where God wanted me to be. I still have that peace.

34

Everybody now – getting involved without getting arrested

"I'd love to get involved in Extinction Rebellion, but I'm not able to get arrested." Have you heard that before? Maybe you have said it yourself.

This movement needs activists who are willing to put their freedom on the line and risk arrest. We need more of them, and we would encourage many more Christians to prayerfully consider if that is a sacrifice they can offer. But not everyone is called to that, and there are many other ways to get involved.

When we watch a film, it's the actors that are in the foreground. It's only when the credits roll that we see how many hundreds or even thousands of crew members it took to bring that movie to the screen. Extinction Rebellion is similar. Those on the front lines are the ones who appear in the news footage, but there are so many important roles behind every action.

We're confident that whatever your skills, whatever your circumstances, there's a place for you in the rebellion!

Taking part in actions

Let's start with direct action, and the fact that thousands of people can be involved in an action without getting arrested. There are no guarantees with civil disobedience, but the majority of people who find themselves in a police van have chosen to be there. They are locked or glued on to something, have done something visibly illegal, or have chosen to disregard a police request to move. We can't vouch for the actions of any single police officer, but they will almost always give a warning and a final warning before making an arrest.

Sometimes the difference between arrestable and non-arrestable participation is a matter of timing, perhaps being among those who hold a road, rather than those who step out first to stop the traffic. If you keep an eye on police movements, you can often tell when it is time to leave. Location matters too: police might move faster at an airport than a roadblock, for example. There is strength in numbers of course, and big actions with thousands of people involved present a lower risk.

Protest situations can be disconcerting. "The small experience of stepping out for the first time to participate in an act of civil disobedience felt big to me!" says Ruth Faber from CCA Belgium. "But all of these small acts are what create the powerful ones." If you have any concerns, go with more experienced activists who can help you assess the situation.

"I work in mental health and being arrested could be detrimental to my career," says London-based activist Holly Petersen. "Despite this, I feel secure enough to have been on the front lines at dozens of CCA actions. Protests can look chaotic, but from my experience there is an order to the mass mobilization actions and you usually have to work quite hard to be arrested. That said, if you would rather not be arrested, you do need to have your wits about you and not do anything which may provoke the police."

Supporting roles at actions

Direct action is a team exercise, with many supporting roles. These aren't guaranteed to be 'unarrestable', but if you want to be at the heart of the action, here are some roles you could consider.

First, there is always somebody serving as a *wellbeing officer*. They keep the activists fed and watered, warm and cared for. They look after people's phones and belongings when they are arrested, and make sure they get them back afterwards! They are a reassuring presence, especially when things get stressful. Those serving in this pastoral role wear blue hi-vis vests to show who they are.

A related role is *arrestee support*. Activists are often released in the middle of the night, which can be lonely and disorienting. Arrestee support volunteers stay outside the police station and wait, ready with a hug and a lift home. The hours are long and sleepless, and you don't get your photo in the paper! In other words, it's a sacrificial role that almost no one will see. We do it for love. CCA member Liz, from Exeter, sees her role here: "I don't currently feel I am in the right headspace to be arrested personally, but I want to be part of the essential support network around those who are putting themselves at risk of arrest or being arrested for this cause."

Legal observers are also present at major actions. Identified by their orange vests, they make a note of police movements, record arrests and take names for witness statements. They remind people of their rights, and they keep both police and rebels on their best behaviour. However, they don't intervene, and must remain neutral and not visibly part of the protest. *Police liaison* officers are sometimes used as well, especially on major actions. Extinction Rebellion provides training for this specialist role.

Performance roles at protests

If you've spent any time with Extinction Rebellion, you will have seen the role of performers at its protests. They bring colour and energy, at other times real pathos and emotion. There are a whole range of different kinds of performance, from the musicians and poets that take to the stage, to the drummers or choirs of singers that move from site to site. There are clowns, dancers and actors, and these skills can provide a way into the movement. As CCA supporter Rachel Webbley observed, it was "an invitation to sing in a choir that made me able to see myself attending London actions with Christian Climate Action."

Perhaps you might want to join one of the established performance groups such as the Red Rebels Brigade. Jan Smith, a CCA member from Norfolk, recognizes the Red Rebels as "prophetic and powerful, illustrating the loss and grief I feel about the suffering and death of people and the natural world." Or you might develop your own piece of street theatre, such as the mock wedding Christian Climate Action held between fossil fuels and the church, calling for divestment.

Ministering at actions

It crosses over with performance to a certain extent, but one of the distinctive things about Christian Climate Action is that we bring elements of worship and spiritual practice into our activism. We can always use *worship leaders* and *liturgists*. There are some suggested songs in the resource section of this book, though you never know what the moment might call for. A social justice campaigner from an independent church in Hampshire recalls seeing "nuns in Trafalgar Square singing 'All you need is love' to de-escalate a potentially hostile situation."

As part of the Faith Bridge programme in October 2019, we had *readers* who read the whole of Revelation from the steps of the National Gallery. *Priests* led us in communion, and on one occasion hosted a baptism service in a paddling pool. Others led regular times of prayer and meditation. Bring your ministry, and experience the prophetic charge of doing it in the street, outside a government ministry, or on an occupied bridge!

Press and media

Getting press attention isn't the only reason to take action, but it's the one that can really push the conversation further. It can take a lot of work to get a story in the papers. A *press officer* might tip off journalists ahead of time, and then provide a first point of call for follow up. They gather quotes from key people, and put together a press release. They might also be running Twitter or Facebook updates, though you might have a dedicated *social media volunteer* on the case. *Spokespersons* are also useful, perhaps following up an action with a radio interview.

Actions are also supported by *photographers*, documenting the event and making sure the press get quality images to accompany any articles. *Videographers* may be on hand too, sometimes providing a live feed, or gathering footage to edit together later. "I've never been able to resist filming whatever is going on, so being part of CCA actions was no different," says Peter Armstrong, who has made several videos for CCA. It isn't always easy being a detached observer, filming while others are arrested. "I felt I was not just watching, but learning – learning from everyone involved what it meant to have one's faith, love and commitment tested to the limit. That's what I tried to bring out as I edited each piece."

Storytelling

The media is one aspect of communications around XR, but there's a broader role for storytelling too. The movement needs *writers* who can describe their experiences for those who weren't there, whether that's articles, email newsletters, or web updates. We need *speakers* who can share stories in churches, local XR meetings, colleges, festivals.

That storytelling can be artistic or musical. Samantha Lindo, a *songwriter* from Bristol, wrote a song about being arrested with CCA and has incorporated it into her performances.

Photography isn't just for the news either, but for sharing stories and encouraging others. As Gillian from North Staffordshire noted, "I found the photographs of specifically Christian events during October incredibly inspirational."

Logistics

Another huge area of behind the scenes support is logistics. A big action can have enormous teams of people working for months at a time on procuring equipment, organising transport, and making sure that everything gets delivered and set up correctly. In the October Rebellion that included organising toilets, buying camping equipment, setting up stages and providing solar panels for power. That requires *electricians* and *engineers*, *drivers* and people who are happy to help out moving things around. Even a small action needs organizing, getting people to where they need to be on time, with the equipment they need.

We ought to include catering here as well. On a big action, there are kitchens on site where people can get food and a hot drink. Large rotas of *cooks, washer uppers* and *potato peelers* keep the food coming. The love and dedication found in the kitchens can be a sign of the kingdom, as Sandra Sutcliffe discovered: "volunteering in the kitchen at the Spring Rebellion in Bristol enabled me to see how the world could be if we all actively cared for each other."

If you're there on a more occasional basis, you can bake a cake, or bring in sandwiches or wraps. Make it vegan so that nobody is excluded.

228

Arts and craft

"Art is a big part of Extinction Rebellion," says CCA member Ceri from Conwy, "and at certain times in its history, it was a big part of the Christian church. Art needs to become part of the fight." That means we need *artists* and *graphic designers*, people who are handy with a *sewing* machine who can run up costumes, or *printers* who can make stencils or t-shirts.

Artwork comes in all sizes. It might be as simple as designing a button badge or a sticker. It might be something considerably larger. For the October Rebellion, "The call went out for a *carpenter* who could make an Ark," says Barbara Keal, "and I volunteered my husband Rich. For the two and a half weeks before the rebellion he worked long days designing and building a beautiful ark with the young people at the school where he teaches woodwork."

Making things can be a lovely way of involving people who can't or won't be there in person. Hundreds of prayer flags were made ahead of the Faith Bridge action, contributed by churches ahead of time.

Extinction Rebellion uses printing at bigger protests and at festivals, and there's always room for more *art workshop facilitators*. These can be powerful. One CCA member, Rev Christopher Maclean, encountered XR at the Greenbelt Festival. He describes how "in a cathartic moment when I printed a piece of fabric with the XR logo, I knew I was committed."

In between actions

The energy within Extinction Rebellion moves in waves around major actions, but there is always plenty happening in between. There is awareness raising and recruitment to do, forming new regional groups, welcoming people into the process. That needs a lot of active *administrators*. There's lots of demand for *trainers* in non-violent direct action too, getting people prepared and confident. *Fundraisers* can generate funds for actions, or to help with transport and legal expenses. *Researchers, planners* and *regional coordinators* will have plenty to work on.

There are specific tasks for after actions as well, as part of the regenerative culture that we seek to practice. People need to process their experiences and recharge, work through things that have been difficult. There's a role here for *counsellors* and good *listeners*. Of course, it's important to celebrate successes as well, so sometimes there's an opportunity for *party planners* and social *organizers* too!

If you're involved in XR it is only a matter of time before you know someone who will have to go to court. If you have relevant expertise you could offer *legal support*, but going along on the day or waiting outside the courtroom can be a huge support.

Prayer and teaching

Civil disobedience is controversial, and one of the things that has helped Christian Climate Action to grow is *church leaders* and *teachers* who speak out in support, or who preach about the Climate Emergency from their pulpits. We want more churches to have the opportunity to play a part in the most important challenge of our generation, and that means pastors and leaders who can teach and train and support the movement.

That's a long list of roles and opportunities. At the end, here's one that we know you can do: *pray*. It is of course the most important thing of all – if that's all you can bring, don't believe for one minute that it is a small or insignificant contribution. The people who pray for Christian activists, and who intercede and mourn for God's creation, are the beating heart of the movement.

Elaine, a third order Franciscan, found her role within XR through prayer, as part of the 'Earth Pilgrimage' walk to London. "I had a deep conviction that I should be involved in XR actions, but as someone with Aspergers, I knew that I could not handle being in the middle of a noisy crowd in a city. God met me where I was, and it was my part to walk prayerfully through the Welsh landscape that is my home, thanking him for its beauty and its generosity, every step deepening my commitment to the need for climate action."

We have put prayer at the end, not because it is an afterthought, but because this is where we begin. As you think about what your role in the Rebellion might be, pray about it. Where is God wanting you to get involved? Where might Christ be calling you? Are you prepared to make sacrifices if necessary? What price would you be willing to pay to truly follow Him?

Pray a dangerous prayer, with courage and honesty, and be open to the answer.

Let us think of ways to
motivate one another
to acts of love.

Hebrews 10:24 (NLT)

35

Supporting XR as a family

CHRISTIAN CLIMATE ACTION

One of the strengths of a non-violent movement is that it welcomes everyone. There is a place for all ages, races and backgrounds. There is a role for young and old alike, for nursing mothers, school children, young professionals, grandparents. This diversity prevents the discussion being dominated by any one group of people, and actively demonstrates unity of purpose. It reflects the fact that the climate crisis is an inter-generational injustice. And as the school strike movement has shown, the leadership of children brings a moral clarity that is hard to ignore.

Involving children can be a challenge. There is a balance to strike in being open and honest about the climate emergency, and burdening young minds with the worries of the world. Youth activism has to be done sensitively, and every parent will know their own children and what they are ready for. Yet we welcome families as a movement, and we hope that yours will find a place in the rebellion.

Baby rebels

As many a young parent has discovered, babies can prove highly portable and there are plenty of pushchairs and slings at XR actions. "I took my son, aged nine months at the time, to the April Rebellion" says XR activist Sam Willis. "We joined the samba band with him in the sling, and he joined in and loved it. We want to bring him up with a sense of his own political agency, and this felt like a safe and fun way to get started."

Mothers and babies have taken part in their own dedicated actions too. Laura Condell, a mum from City Life Church in Luton, describes taking part in a nurse-in during October 2019: "Nearing Westminster, it was impossible not to catch the sense of joy as the crowd of infants, mums, dads and grandparents slowly built. Random strangers expressed thanks for us being there, which seemed a little strange. To me, it felt such a tiny effort compared to those who had taken time off work and risked arrest for a fortnight to make this message heard, let alone those who risk and give their lives the world over."

"It was in the calm beauty of the nurse-in, with all of us sitting down to feed our babies in the road near Downing Street, that the significance of what we were doing began to reveal itself. The power of the demonstration lay in bringing these tender, precious, tiny lives to momentarily take centre stage. We find the fragility and vulnerability of early infancy, and of the world's ecosystems, equally hard to bear in our minds. There, for a short time, both had full attention."

"I've found becoming aware of the climate crisis hard at times, grieving the loss of the certainty I thought I knew, grieving a sense of the world which I won't be able to hand on to my children. But I draw comfort from the love and solidarity which was tangible in that moment on that day, as we sang and nursed (and wrangled) our children. Maybe, if we can open our hearts to respond to God's call at this time, we might just hand them on a better one."

Family friendly actions

Larger Extinction Rebellion actions often have a child-friendly area that is promoted in advance. Sometimes this is just a safe space where families can base themselves for the day, sometimes there are craft activities or games. XR Families also plan actions, such as the Families Sing for the Climate event during the October Rebellion, or a 'love-in' at the National Maritime Museum.

"I took my four-year-old to the April Rebellion on a day the XR Families group were doing a family friendly die-in at the National History Museum," says Christian Climate Action member Caroline Harmon. "What struck me is that if you take children to these sorts of actions from an early age they are just going to think it's normal – it's normal to challenge things that are wrong, and that need changing."

When XR occupied five bridges in London in 2018, parents met online and agreed to create a child friendly space on Blackfriars Bridge. "Blackfriars Bridge was the first XR event I went to, and my favourite one so far," says Zach, aged 8. "We helped to block the bridge, and we drew all sorts of messages on the pavement with chalk. I like Extinction Rebellion actions because they're fun, but we know that it's also making the world a better place."

"I liked making badges," adds Eden, age 6. "We went on the train to Blackfriars, and we took our badge machine.

We made badges and gave them out to people. They had animals on them, or the Extinction Rebellion symbol."

"We've brought the kids along to a few actions now" says Zach and Eden's dad. "Usually we're just part of the crowd, but I've also seen situations where children really changed the dynamic. I brought them to a local consultation on airport expansion that XR had been protesting at. We turned up with soft toy animals holding little protest signs, and sat them on the table with all the planning proposal documents. It was like a mini occupation. It was silly really, but the whole room fell silent. One seasoned protestor had tears in her eyes. It was a stark reminder that the decisions being made in that room would affect future generations, and until they arrived nobody was thinking about that."

Honouring the leadership of children

One of the most critical interventions in the climate debate in recent years has been Greta Thunberg and the school strikes. It broke through to adults in a new way, and activated young people around the world. Jo Richards, a rector from Canterbury who took part in the October Rebellion, recognized the role of her children in getting her along. "My 15 year old daughter spent a whole day doing a banner, and gave up her favourite lessons and broke her 100% school attendance record to go out onto the streets of Canterbury – if she can do that, then the least I can do is this." As parents and children learn from each other, there is a real opportunity to "be role models to one another and live the gospel."

Children have not just inspired adults to participate, but also led actions themselves. Elsie Luna, aged 11, has supported Christian Climate Action rebels on actions and outside court. The adults were able to return the favour when Elsie came up with her own idea: "My mother showed me an article about the 100 companies that are most responsible for global carbon emissions. I found out that many have offices in London. I decided to go and find the leaders of the companies, and ask them why they allow these emissions to happen when we know what will happen to life on the planet?" Fellow activists accompanied Elsie from office to office. Not every company was willing to meet, but several did. "I told them what percent of emissions their company is responsible for," says Elsie, "and to please, please have a change of heart, declare a climate emergency, and keep the fossil fuels in the ground."

For families with older children and teenagers, XR Youth is a well-established network that organizes events and actions of its own. Some actions are multi-generational, others exclusively for young people, but there is often a supporting role for adults.

Grandparents of the rebellion

Family participation isn't just about children. Phil Kingston writes about 'climate eldership' elsewhere in this book, and there is a huge role for 'elders' within the movement. XR Grandparents have run their own actions or partnered with others – such as a sing-a-long with XR Kids outside Buckingham Palace in October 2019.

While some grandparents are activists themselves, others have found a supporting role. Sarah Cornick, from South London, joined Christian Climate Action in Trafalgar Square with her five month old son. Her father came along too, which was a meaningful experience for Sarah: "It's been lovely. He doesn't have a background in climate activism, like me, but he does care about it and was just curious to see what is going on. If by that he learns a bit more and gets involved then that's wonderful. That's why it's so important that everyone here is so friendly. This is everybody's world and we want everyone to get involved in whatever way they can."

Another CCA member occupied Waterloo Bridge with three generations of his family. "My parents have been involved with protests before, but not direct action. That sort of crossed a line for them. Then you've got the papers talking about anarchists and disruption, and even the name 'Extinction Rebellion' is quite difficult language. I can see how my parents were wondering what we'd got ourselves into. I was so pleased when they decided to come and join us. As soon as we got to the bridge, I could see their perspective changing. It was vibrant, colourful, welcoming. There was so much joy, and they were swept up in it too. They won't be part of XR themselves, but there's a shared understanding now that is very helpful."

36

Distinctively Christian direct actions?

CHRISTIAN CLIMATE ACTION

In the 1970s, the activist Gene Sharp compiled a list of every non-violent form of protest he had ever come across. The list ran to almost two hundred and included everything from strikes and walk-outs to mock awards, guerrilla theatre or 'collective disappearance'. Since the advent of the internet and social media, activists have thought of several more.

In short, there are many different forms of protest. What do we choose? And as people of faith, are there any actions that would be particularly powerful? There are lots of other stories and perspectives in this book already, and there are no right answers to those questions. Protest is highly contextual, and a lot depends on the audience and what the important messages are. All kinds of creative ideas are possible, but here are some actions that CCA has taken part in, to get you thinking.

Prayer

Prayer is a distinctive feature of Christian Climate Action. As Caroline Harmon says, "We can take any protest action that anyone else is doing, but prayerfully." Prayer precedes, accompanies and follows every CCA action. It can also be a protest in itself.

For example, Caroline shares the story of an early CCA action: "Phil Kingston and Reggie Norton, both in their 80s, wanted to pray outside Downing Street. David Cameron was Prime Minister at the time and had been invited by the head of the UN to go to a climate summit for heads of state. He wasn't going, and they wanted to pray about that. They kneeled in front of the gate, blocking traffic in and out. About ten of us were either side of them, and I was legal observing. It was a much smaller action than our XR events later, but really powerful that our elders wanted to do that for us." David Cameron did attend the summit in the end.

Of course, you may feel moved to pray about an issue like that without turning it into a protest, but praying 'in context' can be meaningful in a different way. "There's a story of a priest in the United States who was arrested at a nuclear base," says CCA's Martin Newell. "They asked what he was doing, and he said he was praying for peace. 'Couldn't you do that anywhere?' they asked, and he said that if somebody is sick and in hospital, the priest will go to them, and maybe lay hands on them. It's important to be where the sickness is."

Or as the American activist and satirical street preacher Reverend Billy puts it, "I find that praying while trespassing has more power, Amen?"

Declarations and confessions

Protest movements, including Extinction Rebellion, regularly use spoken declarations to build common purpose and to be transparent about our intentions. For CCA this is often a moment to explicitly connect our faith and our actions, as it did when leading a litany at the launch of Extinction Rebellion in 2018. "With the help of God's grace," it read, "let us resist and confront evil everywhere we find it" The crowd then repeated the refrain, "We will not comply" as a response to the destruction of creation, a culture of consumerism and the silencing of protest.

Confessions add another angle because they admit our own complicity. This is another distinctive aspect of Christian protest – a humility that recognizes that we are part of the problem too, and we need forgiveness. We are not better than others, nor do we have all the answers.

Vigils

Traditionally, a vigil was a time of purposeful wakefulness, a 'keeping watch'. It might be done as part of a funeral rite or 'wake'. It could be held in expectation, such as an Advent service. The Jewish Passover tradition includes a vigil, held to remember the night before the people of Israel were liberated from slavery – a sleepless night in anticipation of freedom and rescue.

In the context of protest, a vigil can combine many of those meanings – mourning a loss, hoping for change, and bearing witness to injustice. "Others may sleep," says 1 Thessalonians 5.6, "but we should stay awake and be alert."

Among others, CCA has held vigils at fracking sites, government offices, and at the railway terminal of the Heathrow Express.

Scripture reading

Declarations or public readings can also include Scripture. During the October Rebellion the Book of Revelation was read aloud from the steps of the National Gallery. "Revelation was written at a time when there was a sense of crisis, and Christians felt themselves to be right up against it" says Paul Bayes, Bishop of Liverpool and one of the readers. "The writer of Revelation had these visions, some of them hard to understand, but overall there's a sense of urgency, that the risen Jesus moves among the churches and asks us to wake up."

The language of ecological disaster is hard to miss in Revelation: "a third of the earth was burned up, and a third of the trees were burned up, and all green grass was burned up" (Rev 8.7). It revealed the text in new ways to songwriter and activist David Benjamin Blower. "The voices read the disturbing text against the sound of helicopters circling low," he writes, "part of the effort to contain and wear down the sites of civil disobedience. I have always loved the Book of Revelation, and borrowed from it endlessly. And those anti-imperial words never rang so true as they did on Trafalgar Square before those Romanesque pillars of the British Empire."

Worship

In the broader sense of the word, all of CCA's actions are worship – an offering of our bodies as a living sacrifice as Paul describes in Romans 12:1. But sung worship plays an important role in CCA actions too. Communal singing builds togetherness, which is why it's a big part of Extinction Rebellion more generally. For Christians, singing explicitly frames the action as an act of worship, communicating to ourselves and to others that we act from our faith. It's also one of the more immediate ways to identify Christians within the movement: "We are protesting, and we are also a public witness" says Caroline Harmon. "We're honest about who we are."

For the October Rebellion, CCA prepared a songbook, including songs from other faiths alongside Christian hymns. It was used in the march described in the Faith Bridge chapter, where the activists of all faiths and none sang Amazing Grace together as they crossed the Thames.

Margaret Bullit-Jonas, an Episcopal priest and climate activist from Massachusetts, wrote about how risking arrest for the first time led to a breakthrough in understanding. "After years of going to church, after years of celebrating Communion, only now, as I kneel on pavement and face a phalanx of cops, do I understand so clearly that praising God can be an act of political resistance. That worship is an act of human liberation... I feel as defiant as a maple seedling that pushes up through asphalt. It is God I love, and God's green earth. I want to bear witness to that love even in the face of hatred or indifference, even if the cost is great."

Eucharist

There are of course many layers to the central ritual of the Christian faith, the Eucharist. It has featured prominently in CCA actions, sometimes in support of activists, and sometimes as part of the protest. "I presided at the Eucharist on the first night of the October rebellion" says the Reverend Jonnie Parkin, a pioneer priest from Grantham. "We had planned a Eucharist on the bridge but knew this would not happen, so we decided to celebrate together where we were. Recalling both the Passover liberation of God's people and the last supper and death and resurrection of Jesus in that context was powerful, and it transformed my view of the eucharist as a subversive act. I took communion to people involved in the sit-down protests as people all around us were getting arrested. It felt like the upper room and Gethsemane rolled into one. After the service I shared peace with some of the onlooking police officers."

Sister Katrina Alton was among those being arrested. "As a Catholic the eucharistic symbol doesn't only point to the past or the future, but blessing, breaking and sharing our very selves. We prayed and sang, broken. We were dragged away and arrested. We ourselves were shared, using our place of privilege to take that suffering away from the poorest already living the climate emergency."

Other communion services during the October Rebellion was less eventful, but still powerful. "Taking part in the Eucharist in Trafalgar Square with you all felt like the Spirit setting us on fire again" said one participant. Another noted a comment from the stranger standing next to him. "I whispered to him 'I'm not an Anglican, but I love this,' and he replied, 'I'm not a Christian and I love this'."

Baptisms

Along with communion services, the October Rebellion also saw baptisms. Led by the Rev Jon Swales, a dozen activists re-affirmed their baptisms in a paddling pool in Trafalgar Square – a powerfully symbolic act and in some ways a form of street theatre as well. During the baptisms Jon used the sign of the cross, pointing out how "That symbol was there to say 'that's how rebels die', but instead it has become the focus of our hope."

Holly-Anna Petersen was among those reaffirming vows, and she explained that "I wanted to be baptized here because this is the outworking of my faith. Being here, making a stand for God's creation, is part of my worship. Jesus was the ultimate rebel, whose bravery in speaking out against injustice led not only to his arrest but his death."

Other actions

There are many other actions we could include here. The April Rebellion in London saw activists wash each others' feet. CCA has used street theatre, including a mock wedding. Some have taken part in fasting or hunger strikes. CCA in Australia held an action where they processed while carrying a large wooden cross. Then there are the many times that CCA members have blocked buildings or roads, sprayed paint, climbed under or onto vehicles, glued or locked themselves into position, or joined in with protest actions organized by others. Some of them are obstructive, some creative. "The blood of our children action," where symbolic blood was poured in the street, "was the most powerful thing I have ever seen or been involved in," said CCA member and eco-theologian Rachie Ross.

What we choose to do matters, because we want our actions to be strategic and well understood. But how we act is just as important, and it is perhaps there we can see what a distinctively Christian protest act might be. We act prayerfully. We seek God's guidance. We try to do what is right, and seek forgiveness and learning when we get it wrong. We understand that what we do is part of a whole life discipleship, and a call to follow a rebel Jesus. And most importantly of all, "we have to come at the world with a different energy than fear, hate, and anger," as Jamie Stewart from Edinburgh says. "I absolutely believe it is the hard work of faith, hope, and love."

Melanie Nazareth serves
the cup during Eucharist
on the Faith Bridge.

37

How to declare a climate emergency

GREEN CHRISTIAN

One of the key demands for Extinction Rebellion is that people 'tell the truth' about the environmental crisis, and declare a Climate and Ecological Emergency. That's something governments, political parties and councils can do, but also businesses, arts and cultural institutions, interest groups, charities, NGOs and universities.

Churches can declare a climate emergency too, as denominations, parishes or individual congregations. New declarations are being made all the time, such as the one from the provincial synod of the Anglican Church of Southern Africa, which acknowledged that "we face a triple emergency of climate change, biodiversity loss, poverty and inequality."

Pope Francis has stated that we face a climate emergency, and that "we must take action accordingly, in order to avoid perpetrating a brutal act of injustice towards the poor and future generations... The climate crisis requires our decisive action, here and now, and the Church is fully committed to playing her part."

Bishop Andrew Watson declared a climate emergency for the Diocese of Guildford. "As Christians we need lift our voices to join with the growing environmental movement" he said, "to tell the truth about the climate catastrophe, to repent of the behaviours that has caused this emergency, and to prioritize this ministry as an act of sacrificial love to all people, including those yet to be born."

What can your church do?

At Green Christian, we have been calling on the national governing bodies of Churches and denominations to use their moral leadership and spiritual insight to help mobilize change. A declaration of emergency can act as a catalyst for rapid, focused, co-ordinated and equitable action across society, and help individuals make sacrificial choices which reach well beyond the incremental change achieved so far.

However, don't wait for leadership from the top. Start where you are, with your own church, and declare an emergency with your own congregation. If you are active in your denomination, work upwards through local and national levels, using your church's governance structures. Build relationships, keep asking courteously and persistently, and see where it takes you.

John Ranford, an Extinction Rebellion activist from Swindon, did this with the Methodist Church. He grew up in a missionary family in Tuvalu, a small island state that is very vulnerable to climate change, and where people are already migrating to safety. His call that the Methodist Church declare a climate emergency was added to the agenda for a regional synod, where "I did a ten minute speech and it was agreed unanimously. They then referred it up to the national synod. It went through a vetting procedure, and it was put to them with an amendment."

The Methodist Conference officially recognized a climate emergency in summer 2019. "For outsiders, they want to know that the Methodist church is active on the climate" says John. "And it empowers people to act at the local level, and get the climate on the agenda. They can go to their own churches and say 'synod says we should do this.'"

When you make your declaration, you might want to elaborate with an appropriate theological or ethical statement. It will carry more weight if you include recommendations or pledges for action – for example, the declaration from the bishop of Guildford included seven climate-care commitments.

Here is some sample text for a declaration of climate emergency.

We [*add name of governing body*] declare a Climate and Ecological Emergency
We pledge to work with and support our congregations and government, at local and national levels, in tackling this Emergency, and we call on others to do the same.

These are our intentions:

1. **We will tell the Truth**
   Governments and public broadcasters must tell the truth about the Climate and Ecological Emergency, reverse inconsistent policies and communicate the urgency for far-reaching systemic change.

   We will communicate with members of our congregations and support them to discover the truth about the Emergency and the changes that are needed.

2. **We will take Action**
Governments must enact legally binding policy measures to reduce emissions to net zero as soon as possible and to reduce consumption levels.

Acknowledging that for those already suffering any future deadline is too late, we pledge to work towards reducing our emissions to net zero by [insert your chosen deadline here]. We will challenge policies and actions of local and national governments and their agencies, where they do not help to reduce emissions or consumption levels.

We will actively work to imagine and model ways in which our faith and our congregations can enable the planet's resources to be safeguarded and regenerated.

3. **We are committed to Justice**
The emergency has arisen from deeply systemic injustices. Faith communities can imagine and unleash shifts in the ways people relate to one another and the world, in our values and behaviours.

We will do what is possible to enable dialogue and expression among our congregations and the communities they serve about how the Emergency

will affect them and the changes that are needed.
We believe that all truth-telling, action and democratic work must be underpinned by a commitment to justice both within our nation and towards other nations, particularly those who are poorer.

Green Christian is a network that exists to share Green insights with Christians and to offer Christian insights to the wider Green movement. <www.greenchristian.org.uk>.

38

Running grief circles

MELANIE NAZARETH

As the full impact of the climate crisis unfolds, there is a growing recognition of climate psychology. How will we process the grief and loss associated with the collapse of nature and the breakdown of the climate? How can we mourn the depleted future that has been left to our children, and come to terms with our own complicity in destruction?

These are complex questions, but CCA has been developing 'grief circles' as a way of helping people to acknowledge the deep emotions around these issues. These grief circles draw on secular and sacred resources and influences, including the work of Joanna Macy and Francis Weller, adapting and bringing them together in a way that reflects our faith-based traditions of contemplation and lament. We invite you to use this resource in your community, whether in the church or during direct action. They can be profound and life-giving moments.

Preparation: If it is possible, two people should hold this circle, one as facilitator and one as circle-keeper.

Before the circle begins place pieces of paper each with one of the 'grief-gates' written on them around the space, and place around them objects from nature (leaves, bark, seeds, twigs etc). The 'gates' are as follows:

Places That Do Not See Love
What We Are Losing
Fear for the Future
Sorrows of The Earth
Ancestral Grief
Harm We Have Caused

Make a cross of stones in the centre of the circle.

The circle-keeper will quietly welcome people to the circle and will direct latecomers to suitable spaces so that the facilitator can maintain a contemplative and prayerful focus.

The amount of time you give to contemplate at each stage depends on the time you have available, but the key is to keep it calm and unhurried. If time is short leave out the grief-gate work that is on page two. Instead, just ask people to choose one of the nature objects which could be placed on a table or around the room.

Begin the circle with reminder:
This is a respectful space. We listen with our hearts as well as our minds and we recognize that we are all here with concern for the world and for one another. The work we will do here together sometimes brings out strong emotions, and one of the things we do in this circle is acknowledge and accept

these emotions in each other without judgement. It's ok to cry. If you feel the need to take time out or leave please do.

Our personal grief has many gates. We will be reflecting on six of these that might help us connect to our grief about what is happening to the earth, but you may find other experiences and losses coming to mind. That's ok. Just let them be. This process is for you and there is no right or wrong way to experience it.

This is your time to use as you need, but it's not counselling and we are not qualified to offer therapy.

Facilitator explains the grief-gates:

Places That Do Not See Love
These may be very concrete – rubbish dumps and landfill sites; islands of plastic in the ocean. They may be human – those places in other people's being where God's love for Creation is not seen. They may be personal – places in our lives where we shut out love for God's Creation.

What We Are Losing
The breakdown of the climate and the degradation of our environment mean that there are many precious things we will lose. Coastlines are changing. Plants and animals are on the verge of extinction. Perhaps there are personal things that you have lost. Giving up the the things in life we now see cannot be sustained can be hard.

Fear for the Future
Letting go of our expectations and our hopes for the future.

Whatever we do to limit further harm to the planet, our world and the world we leave to future generations will have changed. May be much harder. Less secure. We may be fearful of what the future holds for ourselves and for future generations.

The Sorrows of the World

The suffering of creation. Exploitative practices that harm the earth for profit. Cruel and inhumane treatment of other living creatures. Refugees, many of whom are already climate refugees. Global inequality and poverty multiplied by climate change. The sorrows of the world are so numerous, that maybe we find ourselves overwhelmed.

Ancestral Grief

This can be a tough one to acknowledge. Some of us carry the injustices done to our forefathers and their lands. Some of us bear the burden of inherited privilege gained through the exploitation of others and of the Earth. It may be that we grieve that we have lost our ancestors connection to the Earth.

The Harm We Have Caused

This speaks for itself. We may wish to acknowledge how the way we live our lives is damaging Creation. Our sorrow at our own part in the global crisis.

Facilitator invites everyone to stand up and have a shake out. Facilitator explains:
Slowly meander around the space. Allow yourself time to think on the grief-gates. Which of the gates are you

particularly drawn to? Pick one of the objects from nature from that gate and return to sit with your piece of nature.

As people return the facilitator says:
Let's just sit quietly for a moment and centre ourselves.

When everyone has returned (circle-keeper may have to ask people to return if time is short) facilitator says:
Let's just sit with our grief for a while.

Give yourself time just to be still and present. Sit and listen to the sounds around you but hold yourself apart from them. They are just there, they don't need your attention.

Silence

Facilitator says:
Closing your eyes, focus on your breathing. Don't try to breathe any special way, slow or long. Just feel the breathing as it happens, in and out. Note the accompanying sensations at your nostrils on your upper lip, in your chest and abdomen ... As you notice your breath, note that it happens by itself, without your will, without your deciding each time to inhale or exhale .... It's as though you're being breathed—being breathed by life ... Just as everyone in this circle, everyone in this place, everyone on this planet now, is being breathed by life, sustained in the vast living breathing web of the Creator ... 'For in him we live and move and have our being.' (Acts 17.28)

Silence

Facilitator says:
Breathe deeply. Let your breath out slowly. Allow yourself time to acknowledge your fears for the earth. Allow your sadness, your despair, your worry, your anger to rise, to surface. Allow those feelings to rest in the space you have created. Give yourself permission to acknowledge and be with them.

Pause
And now, with deliberate intention, allow God to enter into the space you have created.

Silence

Facilitator slowly and purposefully says:

Let God hold you as you take a deep breath. Breathe slowly in and out. Visualize your breath as a stream of air. See it flow up through your nose, down through your windpipe and into your lungs. Now from your lungs, take it through your heart. Picture it flowing through your heart and out to reconnect again with the larger web of life.

[Think about the grief-gate that drew you to it,] open your awareness to that suffering that is present in the world. Drop for now your defences and open yourself to your knowledge of that suffering and loss. Let it come as concretely as you can ... images of the wildfires, of flooding, of vast fields of monoculture, of drought, animals and fellow-beings, of thirst, hunger, loss and fear. No need to strain for these images, they are present to you. Just let them surface ... the vast and countless hardships of the groaning earth.

Now breathe in the sorrow like granules on the stream of air, up through your nose, down through your windpipe, through your lungs and heart – and out again into the world ... You are being asked to do nothing for now, but let it pass through your heart ... Be sure that stream flows through and out again; Don't hang on to the images, let the pain and fear go with them ... Surrender yourself silently to God's mercy ...

Silence

Facilitator says:
We have forgotten who we are.
We have alienated ourselves from the unfolding of the cosmos.
We have become estranged from the movements of the earth.
We have turned our backs on the cycles of life.
We have forgotten who we are.

We have sought only our own security.
We have exploited simply for our own ends.
We have distorted our knowledge.
We have abused our power.
We have forgotten who we are.

Now the land is barren.
And the waters are poisoned.

And the air is polluted.
We have forgotten who we are.

Now the forests are dying.
And the creatures are disappearing.
And humans are despairing.
We have forgotten who we are.

Pause

Facilitator says:
Breathe deeply now and allow hope to enter the space and be with you.

Pause

Facilitator continues:
Our hope is not an optimism that we can avoid what we have brought upon ourselves nor that things can return to how they were.

Our hope is in the depths of our sorrow and our repentance.

Our hope is a life-giving process in which we can reconnect the Earth with our God-given purpose.

Pause

Facilitator says:
Take the piece of nature that you chose [from your grief gate] in your hands. Let your fingertips stroke its texture, feel it against your skin.

Pause

Facilitator says:
Open your eyes and look closely at the detail, marvelling at its intricacies. Feel your connection to your object and your object's connection to the Earth and feel this connection flowing through you to the Earth and through the Earth to you. Remember this connection.

You are rooted in God's incredible creation.

Pause

Facilitator says:
We ask forgiveness.
We ask for the gift of remembering.
We ask for the strength to change.
We have forgotten who we are.
(*We Have Forgotten Who We Are – Prayer from from the U.N. Environmental Sabbath Programme*)

Silence

Facilitator says:
When you are ready, bring your object and place it on the cross of stones. If you feel comfortable you may wish to say a word or two that expresses your grief or you may just stay silent.

Circle-keeper may need to lead the way. When everyone has placed their object and returned to their seat, circle-keeper may light candles and place them around the cross.

The Hands

Facilitator says:
Together we have laid down fears and sorrows, together let us go forward.

If you are comfortable to do so take the hand of the person next to you.

Draw strength and hope in each other.

When everyone who wishes to has joined hands, facilitator says:

If you feel comfortable, close your eyes

We join with the Earth and with each other
To bring new life to the land
To restore the waters
To refresh the air

We join with the Earth and with each other
To renew the forests
To care for the plants
To protect the creatures

We join with the Earth and with each other
To celebrate the seas
To rejoice in the sunlight
To sing the song of the stars
We join with the Earth and with each other
To recreate the human community
To promote justice and peace

To remember our children

We join with the Earth and with each other
We join together as many and diverse expressions
Of one loving mystery: for the healing of the earth and the
renewal of all life.
(*We join with the Earth – Prayer from from the U.N.
Environmental Sabbath Programme*)

Silence

**Facilitator invites everyone to unlink hands. Facilitator
says:**
"Though the mountains be shaken
    and the hills be removed,
yet my unfailing love for you will not be shaken
    nor my covenant of peace be removed,"
    says the Lord, who has compassion on you. (Isaiah 54:10)

Silence

*Facilitator ends the circle with some acknowledgement that
the time together has ended. The nicest way is to say that in
their own time, when they are ready people can return to the
outside world, but time might not allow for this.*

*Melanie Nazareth is a part of CCA and XR; she worships within an
Anglican tradition and is presently doing an MA in Theology, Ecology
and Ethics.*

39

Liturgies, declarations, prayers and songs

When taking action, there is often a moment to make a declaration or to lead a group in a liturgy. These can be powerful moments of lament, confession, or of shared purpose. They can also be a moment of witness, and one of the best opportunities for explicitly connecting our actions to our faith. When leading a moment like this with a large crowd, there may be some who have never taken part in a Christian liturgy or in prayer. Some of the prayers and liturgies are collected here.

Singing can also play an important role. Singing hymns grounds our actions in worship. It builds community and encouragement. When we sing well-known hymns, it gives people a chance to join us in solidarity, whether or not they share our faith. So we've gathered together a few suggestions for relevant songs below.

Camps or occupations can spread over several days, in which case there is the opportunity for introducing a rhythm of prayer to the assembled community. For example, on the Faith Bridge in Westminster, we used a version of the Franciscan Hours. A variety of prayers and liturgies for different contexts are included below.

A communal prayer for climate grief

Creatively inspired by Hannah Malcolm

Instructions to Read

We are living in a climate emergency – which is already having devastating effects on the poorest communities around the world. However, this is something which at some point will affect us all. Scientists have said that if we don't act now, we could have climate breakdown as early as 2030–40.

When faced with something which is so difficult for us to comprehend it can be easy to go into a state of denial – to procrastinate or to downplay the intensity of the situation.

As a Church we are called to be faithful – even in these desperate times. However, in order to do this, we first need to emotionally process what we are facing. Therefore, we are going to take this time to grieve with God all that we have lost and all the we stand to lose through climate breakdown. Only by doing that can we come to understand what hope might look like and how we can take action to make that a reality.

We are going to have 5 minutes of silence, where we will write down on a piece of paper all the emotions this climate crisis stirs in us. Perhaps you want to confess something. Perhaps you are feeling, angry, hopeful, sad, or despairing. Perhaps something happened recently that made you feel afraid. Feel free to be as vague or as specific as you would like. Please don't write your name on this paper, so that each contribution can remain anonymous. I will then collect them all together and read some of them out in a communal prayer where we ask God to comfort and mourn with us in this difficult time.

Iinstructions for Gathering Reflections on Climate Grief
- Hand out a small piece of paper and a pen to each member of the congregation.
- Allow five minutes of silence for people to write down a short reflection on how they feel in relation to the climate crisis.
- Collect in the pieces of paper.
- Find around 10 pieces of paper to read out in prayer.

Prayer
Merciful God, we believe that you uphold and sustain all that you have made, while also lovingly giving us the freedom to live in relationship with the rest of creation.

We ask your forgiveness for the ways we have abused that freedom, through what we have done and what we have left undone. We bring our lament and our longing for a renewed earth to you now:

(Read out some messages on the pieces of paper)

As your people, we come to you weary and heavy laden. We ask that you would make our grief holy in your eyes.

After a moment of silence, I will say: 'Lord, have mercy'. I ask that you also respond by saying: 'Lord, have mercy'. *Moment of silence*

Leader: Lord, have mercy.

Everyone: Lord, have mercy.

Amen

Confession and lament

Rev Jon Swales

This confession and lament was used with an opportunity to light candles, and the moment for that is indicated below. It can be used without that as well.

Leader:
Climate change is real.
Desmond Tutu said: 'Twenty-five years ago people could be excused for not knowing much, or doing much, about climate change. Today we have no excuse."
Lord Have Mercy,

> All: Christ have mercy

Climate change is happening now.
The Poorest Countries are the most vulnerable.
Lord have Mercy,

> All: Christ have mercy.

Climate change requires extensive and sustained action to prevent the unfolding of a disaster of apocalyptic proportions.
Lord have Mercy,

> All: Christ have mercy.

(Leader kneels and faces away from congregation)

Leader:

Almighty God,
we have sinned against you
and against our neighbour
in thought and word and deed,
through negligence, through consumerism,
by being caught up and complicit in economic systems and
lifestyles which bring destruction.
In your mercy
forgive what we have been,
help us to amend what we are,
and direct what we shall be;
that we may do justly,
love mercy,
and walk humbly with you, our God.

All: Amen.

(Candles: People are asked to come and light candles at the
front.)

Lamentations 5.15–22 (ESV)
The joy of our hearts has ceased; our dancing has been
turned to mourning. The crown has fallen from our head;
woe to us, for we have sinned! For this our heart has become
sick, for these things our eyes have grown dim, for Mount
Zion which lies desolate; jackals prowl over it. But you, O
Lord, reign forever; your throne endures to all generations.
Why do you forget us forever, why do you forsake us for so

many days? Restore us to yourself, O Lord, that we may be restored!

Wait 90 Secs

Habakkuk 3.17–18 (ESV)
Though the fig tree should not blossom, nor fruit be on the vines, the produce of the olive fail and the fields yield no food, the flock be cut off from the fold and there be no herd in the stalls, yet I will rejoice in the Lord; I will take joy in the God of my salvation.

Wait 90 Secs

Romans 8.38–39 (TNIV)
For I am convinced that neither death nor life, neither angels nor demons, neither the present nor the future, nor any powers, neither height nor depth, nor anything else in all creation, will be able to separate us from the love of God that is in Christ Jesus our Lord.

Our Father, who art in heaven,
Hallowed be thy Name.
Thy Kingdom come.
Thy will be done on earth,
As it is in heaven.
Give us this day our daily bread.
And forgive us our trespasses,
As we forgive those who trespass against us.
And lead us not into temptation,

But deliver us from evil.
For thine is the kingdom,
for ever and ever,
Amen.

A prayer for Extinction Rebellion

Helen Burnett

This is a prayer for those taking actions. When it was first used, it referred to actions in London and that is the version used here. If you are praying for actions elsewhere, obviously feel free to adapt!

Truth telling God,
weave a thread of love and courage
among those who stand for creation this week.
May they know the sound of your voice in all they do.
May your love echo across the streets of London
so that the sap of change can rise in the mess of the city,
and seep into the corridors of power
to bring the dawning of a new day where the web of life is
sanctified, renewed and replenished.
Amen, Lord have Mercy.

Litany for the earth

Fran Pratt

Begin with this reading from the first half of Psalm 24 (NLT).

The earth is the Lord's, and everything in it.
The world and all its people belong to him.
For he laid the earth's foundation on the seas
and built it on the ocean depths.
Who may climb the mountain of the Lord?
Who may stand in his holy place?
Only those whose hands and hearts are pure,
who do not worship idols
and never tell lies.
They will receive the Lord's blessing
and have a right relationship with God their savior.
Such people may seek you
and worship in your presence, O God of Jacob.
Psalm 24.1–6

God, we lament the destruction that has been done
That we have permitted to be done
By our silence and inaction
And by our direct action
To the Earth – Your creation.
Forgive us, Oh God.

Even now we realize that our home
Is suffering
Its inhabitants are suffering

The Hands

From lack of clean water and air
    Lack of life-giving nourishment
    Lack of safe habitat.

Help us to become aware
    Of the needs of humanity,
    Of the needs of generations to come,
    Of the needs of soil and creatures.
We acknowledge that we have a chance:
    To choose peace over profit
    To choose activity over complacency
    To choose a Greater Good over today's convenience.
Arouse in us a new compassion,
A new willingness to change,
A new excitement to foster community,
A new faith in the abundance of your Kingdom.
A new zeal for establishing the Peace and Justice of God,
A new desire to set the Earth to rights
A new understanding of the connectedness of all things,
A new appreciation of the gift of Earth.
Amen

Litany of Resistance

Adapted from Jim Loney

This litany was written by Jim Loney for the Christian Peacemaker Teams, during the Gulf War. Its powerful rejection of the forces of destruction and commitment to peace are inspiring, and we have adapted it for use in a climate context.

One: Blessed are the poor
All: For theirs is the kingdom of God
One: Blessed are they who mourn now
All: For they will be comforted
One: Blessed are the meek
All: For they will inherit the earth
One: Blessed are they who hunger and thirst for justice
All: For they will be satisfied
One: Blessed are the merciful
All: For they will be shown mercy
One: Blessed are the pure in heart
All: For they will see God
One: Blessed are the peacemakers
All: For they will be called the daughters and sons of God
One: Blessed are they who are persecuted because of righteousness
All: For theirs is the kingdom of heaven.
One: Deliver us, O God
All: Guide our feet to live lightly on your Earth
One: In humility, we ask
All: Guide our feet to live lightly on our Sister Earth
One: In humility, we ask

All: Guide our feet to live lightly on our Mother Earth
One: Lamb of God, you take away the sins of the world
All: Have mercy on us
One: Lamb of God, you take away the sins of the world
All: Free us from the bondage of sin and death
One: Lamb of God, you take away the sins of the world
All: Hear our prayer. That we may live lightly on your Earth.
One: For our denial of our need for repentance from fossil fuels
All: Forgive us for we know not what we do
One: For the scandal of our dependence on fossil fuels
All: Forgive us for we know not what we do
One: For the scandal of billions wasted on fossil fuels
All: Forgive us for we know not what we do
One: For our lack of love for your creation
All: Forgive us for we know not what we do
One: For the global oil and gas companies
All: Forgive us for we know not what we do
One: For the coal mining companies
All: Forgive us for we know not what we do
One: For our scorched and blackened earth
All: Forgive us for we know not what we do
One: For the men and women training to extract and burn fossil fuels
All: Forgive us for we know not what we do
One: For the scientists and researchers finding new ways to extract and use fossil fuels
All: Forgive us for we know not what we do
One: For all the advertisers and all men and women who

perpetuate our delusion
All: Forgive us for we know not what we do

One: Deliver us, O God
All: Guide our feet to live lightly on your Earth
One: In humility, we ask
All: Hear our prayer. That we may live lightly on your Earth.
One: From the arrogance of power
All: Deliver us
One: From the poverty of materialism
All: Deliver us
One: From the tyranny of economic growth
All: Deliver us
One: From the ugliness of extractivism
All: Deliver us
One: From the politics of hypocrisy
All: Deliver us
One: From the hysteria of nationalism
All: Deliver us
One: From the cancer of addiction
All: Deliver us
One: From the seduction of wealth
All: Deliver us
One: From the addiction of control
All: Deliver us
One: From the avarice of imperialism
All: Deliver us
One: From the idolatry of economic growth
All: Deliver us

One: From the despair of fatalism
All: Deliver us
One: From the violence of despair
All: Deliver us
One: From despair and denial
All: Deliver us
One: From denial and spiritual death
All: Deliver us
One: From the death of your gift of glorious diversity of life
All: Deliver us
One: From the madness of avarice
All: Deliver us
One: From the fears of status anxiety
All: Deliver us
One: From the blasphemy of status seeking
All: Deliver us
One: From the demonic waste of our addiction to growth
All: Deliver us
One: Deliver us, O God
All: Guide our feet to live lightly on the earth
One: In humility, we ask
All: Hear our prayer. That we may live lightly on the earth
One: Obedience to God comes before obedience to human authority
All: Render unto Caesar what is Caesar's and unto God what is God's
One: Let your will be done, not mine
All: With the help of God's grace
One: Let us resist and confront evil everywhere we find it

All: With the help of God's grace
One: With the killing of God's creation
All: We will not comply
One: With the normalization of addictive consumerism
All: We will not comply
One: With the propaganda of consumerism
All: We will not comply
One: With the rape of the earth
All: We will not comply
One: With the silencing of protest
All: We will not comply
One: With laws that betray life
All: We will not comply
One: With profit before planet
All: We will not comply
One: With the culture of consumerism
All: We will not comply
One: With the violating of our earth
All: We will not comply
One: With the destruction of peoples
All: We will not comply
One: With the building of walls to keep out climate refugees
All: We will not comply
One: With governments that are blind to the sanctity of life
All: We will not comply
One: With economic structures that impoverish and dehumanize
All: We will not comply

One: With the manipulation and control of public information
All: We will not comply
One: With economics that ignore the gift of God's earth
All: We will not comply
One: With economics that neglect the care of God's earth
All: We will not comply
One: With the perpetuation of violence against God's earth
All: We will not comply
One: With structures that divide rich from poor
All: We will not comply
One: With the hypocrisy of political maneuvering
All: We will not comply
One: With the help of God's grace
All: We will struggle for justice
One: With the compassion of Christ
All: We will stand for what is true
One: With God's abiding kindness
All: We will love even our enemies
One: With the love of Christ
All: We will resist all evil
One: With God's unending faithfulness
All: We will work to build the beloved community
One: With Christ's passionate love
All: We will carry the cross
One: With God's overwhelming goodness
All: We will walk as stewards of creation
One: With Christ's fervent conviction
All: We will labour for truth

One: With God's infinite mercy
All: We will live in solidarity with all people and all life
One: In the end there are three things that last
All: Faith, hope and love, and the greatest of these is love
One: Let us abide in God's love
All: Thanks be to God.

The Joy in Enough Confession

Jeremy Williams

This confession considers the climate crisis in the context of consumerism and an unjust economic system. It was written for the Joy in Enough campaign, and lends itself to use in protests in shopping centres and retail districts.

Our climate is changing, and we are changing it. We confess our carbon footprints, our failure to consider the consequences of our actions, our slowness to react. We are sorry for all the times we knew the right thing to do, but chose convenience.

Your earth is exploited, and we are complicit in its exploitation. Species are lost, soil erodes, fish stocks decline, resources dwindle. We confess that many of us have taken too much, and not considered the needs of future generations.

We have become consumers. We have turned a blind eye to greed. We confess our hunger for more, and our failure to appreciate what we already have. We live in a time of unparalleled luxury, and we are sorry that we have not been more grateful.

The poor are left behind, even in this age of plenty. Human rights are pushed aside for profit. Wealth accumulates for the rich while the poorest still do not have what they need. We confess our apathy to injustice, and our haste in judging others.

This is not who you made us to be. We have not been good caretakers of your garden Earth. We have not loved our neighbours. Forgive us, creator God.

Forgive us. Renew us. Inspire us.

And in your strength, God, we declare:

- Enough climate change: help us to take responsibility. Give us the wisdom to live appropriately, the urgency to act, and the courage to make changes. Give us the voice to call for change from our leaders, and the perseverance to keep asking.
- Enough consumerism: give us what we need, God our provider. Then help us to find satisfaction and contentment. Help us to be grateful and generous.
- Enough inequality: nobody should be left behind. You care for the poor, and we want to follow your example. Make your church a living example of equity and inclusion, and a powerful advocate for justice and sharing.

We thank you for your kindness and your mercy. We look to your promise of restoration, and we move forward. Give us the strength to speak and to act – not out of guilt or duty, for we are forgiven and we are loved. Instead, we speak and act out of joy:

- joy in the living hope of knowing you
- joy in serving each other
- joy in the beauty and diversity of creation, your gift to us
- joy in your provision and your care – joy in enough

Songs for climate protest

These hymns are fairly well known and can often get people joining in, even if the words are not available. There are lots of options, but here are some that we turn to:

Amazing grace
All creatures of our God and king
I the Lord of Sea and Sky
How Great Thou Art
Christ be our Light
This Little Light of Mine
We shall not be moved
As I Went Down to the River to Pray
We are Marching in the Light of God

More songs and chants are available in the Christian Climate Action songbook, available to download at <www.christianclimateaction.org/resources>. As it was produced for the Faith Bridge, it includes some beautiful creation themed songs from other faiths.

Prayer for climate solidarity
Christian Aid

Creator God who has made the earth
The oceans and the rivers teeming with life
The heartbeats of creatures large and small
We thank you for giving us this wonderous ecosystem

Creator God who calls us as stewards
We confess that we have neglected our duty
That human choices are robbing the planet of its life
That climate breakdown has been at our hands

Lord of justice who longs for equality on the earth
We stand with earth defenders and climate strikers
Around the world we join their call for climate justice
May our actions and our words speak life in all its fullness.
Amen

Solemn Intention Statement
Extinction Rebellion

The Solemn Intention Statement is often read before or during an Extinction Rebellion action, and morning and afternoon at each protest site. It serves to ground the action in what matters most, and remind activists why they are there.

Let's take a moment, this moment, to consider why we are here.

Let's remember our love, for this beautiful planet that feeds, nourishes and sustains us. Let's remember our love for the whole world of humanity in all corners of the world.

Let's recollect our sincere desire to protect all this, for ourselves, for all living beings, and for generations to come.

As we act today, may we find the courage to bring a sense of peace, love and appreciation to everyone we encounter, to every word we speak and to every action we make. We are here for all of us.

Notes

Billy Graham quote from *Approaching Hoofbeats: The Four Horsemen of the Apocalypse*, Billy Graham, 1983, Word Books.

Chapter 1

"One cannot also carry stones or clubs at the same time..." The following paragraphs are taken from a leaflet from Nikolai Kirche.

"As I write in Just Living... " see R. Valerio, *Just Living: faith and community in an age of consumerism*, 108.

"A recent report gives us three years..." For background references to these stats, see, R. Valerio, *Saying Yes to Life: the Archbishop of Canterbury's Lent Book 2020*.

Pope Francis quote from *Laudato Si*, The Vatican, 2015. Point 23.

Chapter 2

For more details see <www.christianclimateaction.org>.

Chapter 3

Gandhi quote from M. K. Gandhi, *An Autobiography OR The Story of My Experiments with Truth*, Ahmedabad: Navajivan Publishing House, First Edition – 1927, Reprint, 2011. p.63.

Walter Wink material from *The Powers that Be: Theology for a New Millennium*, Walter Wink, 1998, pp. 98–111.

List of terms: *Violence and Nonviolence in South Africa: Jesus' Third Way*, Walter Wink, 1987, pp. 22–23.

Chapter 4

I am indebted to Candice Delmas's 'Civil Disobedience',

Philosophy Compass, 11/11 (2016) for much of this chapter; the reader is directed there for a sophisticated overview.

"In full public light…" quotes from Extinction Rebellion: <https://rebellion.earth/the-truth/about-us/>

"Many texts in the Bible…" In the New Testament alone this includes: Matt 10.1 –20, Matt 22.15–22 (and parallels), Matt 23 (and parallels), Matt 26.51–55 (and parallels), Mark 11.15–17 (and parallels), Luke 13.31–32, John 6.14–15, Acts 5.29–32, Rom 13.1–7, and many more.

Calvin quotes from John Calvin, *Commentary on Romans*, p. 417 (Grand Rapids, MI: Christian Classics Ethereal Library).

"…as far back as Irenaeus." Victor Vasquez, Studies in the *Rezeptionsgeschichte* of Romans 13.1–7, p. 172 (Goettingen: V&R Unipress, 20112).

"this resistance should be non-violent…" An accessible overview which makes this case from a variety of perspectives is the edited volume by Tripp York and Justin Barringer, *A Faith Not Worth Fighting For*.

Global warming facts drawn from <https://climateactiontracker.org/global/temperatures/>, 2019.

"This level of warming would be devastating." For details on just how devastating these temperature rises would be, see David Wallace-Wells, *The Uninhabitable Earth* (Penguin, 2019) and Jeff Goodell, *The Water Will Come* (Black Inc., 2018).

David Attenborough said "The collapse of our civilizations … is on the horizon" at COP24 in 2018.

Chapter 6

Stories drawn from interviews with Jeremy Williams, and a survey run on the Christian Climate Action website.

Notes

Chapter 7

The Sword and the Plowshare as Tools of Tikkun Olam, Rabbi Arthur Waskow, The Shalom Centre. <https://theshalomcenter.org/node/216>.

"Jeremiah is a prophet for our times…" For a good short survey of Jeremiah and the political background, see <preachingsource.com/journal/jeremiah-the-man-and-his-times/>.

NKJV – verse reference taken from the New King James Version, Thomas Nelson, 1982.

Chapter 8

Martin Luther King's 'Letter from a Birmingham Jail' is available in its full and original form from The King Institute at Stanford University. <https://kinginstitute.stanford.edu/king-papers/documents/letter-birmingham-jail>.

Chapter 9

"While experiencing life under military dictatorship" – Walter Wink, Preface to *Naming The Powers, The Language of Power in the New Testament* (Philadelphia: Fortress Press, 1984).

Quotes from Wink, *Naming The Powers*, p.5, 82.

"Where wills unite…" Dietrich Bonheoffer, *Dietrich Bonheoffer Works, Volume 1: Sanctorum Communio*, ed. by Clifford J. Green trans. Reinhard Krauss & Nancy Lukens (Minneapolis, MN: Fortress Press, 2009) pp. 98–99.

"Western imperialism and corporate capitalism…" William Dalrymple, *The Anarchy, The Relentless Rise of the East India Company* (London: Bloomsbury, 2019) pp. 396–7.

"The ideology of our age…" Walter Brueggemann, *Hopeful Imagination, Prophetic Voices in Exile* (Philadelphia: Fortress Press, 1896) p. 29.

"Largely ignore the institutional sources…" Walter Wink,

Engaging the Powers, Discernment and Resistance in a World of Domination (Minneapolis: Fortress Press, 1992) p. 314.

NRSV – verse taken from the New Revised Standard Version. Division of Christian Education of the National Council of the Churches of Christ in the United States of America, 1989.

"When a particular Power becomes idolatrous..." Wink, *Naming The Powers*, p.5.

"The powers are good..." Wink, *Engaging the Powers*, p. 10.

"One shouldn't underestimate..." Raj Patel, *The Value of Nothing, How to Reshape Market Society and Redefine Democracy* (London: Portobello Books, 2009) p. 176.

Chapter 10

Originally preached as three sermons at St George's Leeds, June 2019 <https://www.stgeorgesleeds.org.uk/talks-all/2019/06/16/1030>.

ESV – Bible verse taken from the English Standard Version, Crossway Bibles, 2001.

Chapter 11

Interview with Jeremy Williams, October 2019.

Leon Sealey-Huggins' work can be explored in detail at <https://warwick.ac.uk/fac/arts/schoolforcross-facultystudies/gsd/aboutus/peoplenew/drleonsealeyhuggins/>.

Chapter 12

For more on why non-violent uprisings are more successful, see *Why Civil Resistance Works: The Strategic Logic of Nonviolent Conflict*, Erica Chenoweth and Maria J. Stephan, 2011.

The non-violent resistance strategy of Jesus, see *Binding the Strong Man: A Political Reading of Mark's Story of Jesus*, Ched Myers, 1988; *Jesus before Christianity*, Albert Nolan, 1976; Dorothy

Day, founder of the 'Catholic Worker' movement, e.g. *Selected Writings*, edited by Robert Elsberg; *Engaging the Powers*, Walter Wink, 1992.

"The moral equivalent of war" is from an essay by William James, 1910. It was later quoted by president Jimmy Carter to describe the energy crisis.

"Research by the movement's founders" – the strategy is distilled in Roger Hallam's *How to Win: Successful Procedures and Mechanisms for Radical Campaign Groups*. <https:// radicalthinktank.files.wordpress.com/2015/12/how-to-win-10-15.pdf>.

The church in Apartheid, see *Violence and Nonviolence in South Africa: Jesus' Third Way*, Walter Wink, 1987.

Chapter 13

First published as 'The fullness thereof – how indigenous perspectives offer hope to a besieged planet', *Sojourners Magazine*, May 2019 <https://sojo.net/magazine/may-2019/ fullness-thereof>.

Reprinted with permission, (800) 714–7474, <www.sojo.net>.

NLT – verse taken from the New Living Translation, Tyndale House, 2015.

Chapter 14

"Deadline-ism" – For a range of perspectives on this, see the special edition of WIREs *Climate Change*, Vol. 11, Issue 1, January/February 2020 e61, edited by Mike Hulme, "Is It Too Late (to stop dangerous climate change)?"

"… since colonial times." See Hayward et al., "It's Not 'Too Late': Learning From Pacific Small Island Developing States in a Warming World", WIREs *Climate Change*, July 2019.

ESV – Bible verse taken from the English Standard Version.

Jem Bendell, *Deep Adaptation: a Map for Navigating Climate Tragedy*, July 27th, 2018 <http://lifeworth.com/deepadaptation.pdf>.

Chapter 15
Written and first performed during the October Rebellion, London, 2019.

Chapter 18
"The more we can acknowledge openly" – see Renee Lertzman's work on ecological grief at <www.reneelertzman.com>.

'The Quickening of John the Baptist' first appeared in the collection *The Tears of the Blind Lions* (New Directions, 1949).

'East Coker', by T S Eliot, was the second poem in his *Four Quartets* (London, Faber and Faber: 1940).

Stages of grief – see *On Death and Dying*, Elisabeth Kübler-Ross (Routledge, 2008).

Climate Psychology Alliance, see <www.climatepsychology alliance.org>.

Chapter 19
First published at <www.dearearth.co.uk>, where Satya writes regular letters to the planet.

Joanna Macy quote from *World as Lover, World as Self* (Berkeley, CA: Parallax Press, 1991), p. 187.

Chapter 20
Chapter based on an interview with Jeremy Williams.
For more on the Passionists see <www.passionists-uk.org>.

Chapter 21
Chapter based on an interview with Jeremy Williams.

Notes

Chapter 22

First written for <www.christianclimateaction.org>.

Chapter 23

Adapted from *The Comforting Whirlwind – God, Job and the Scale of Creation*, Bill McKibben (Cowley Publications, 2005).

This chapter uses quotes from the Stephen Mitchell translation of Job: *The Book of Job*, Stephen Mitchell (HarperCollins, 1992).

Wangari Maathai quote from *Unbowed: A Memoir* (Anchor, 2008).

Chapter 24

This poem was originally written as three tweets @vanessa_vash.

Chapter 25

This article first appeared in Modern Church's newsletter *Signs of the Times*, and is republished with permission.

Chapter 27

"Alone in a world of wounds", Aldo Leopold, *Round River* (1966 edition), p. 67.

"…to walk in the woods in winter", Bill McKibben, *The End of Nature* (1989), p. 211.

Mary Heglar's 'climate visions' described in the *No Place Like Home* podcast, episode 33 <https://podcasts.apple.com/us/podcast/33-all-climate-feels-season-finale-mary-annaïse-heglar/id1158028749?i=1000441855601>.

Planetary Hospice concept from Zhiwa Woodbury, *Planetary Hospice: Rebirthing Planet Earth* (2014). Quote from p. 6.

"… life as we have come to know it," Zhiwa Woodbury unpacks this particular phrasing further in *Principles of Planetary Hospice* (2014), p. 5. Quote from p. 10.

"It is absolutely possible to prepare…", Heglar, M., 'Home is Always

Worth It' <https://medium.com/@maryheglar/home-is-always-worth-it-d2821634dcd9>.

Chapter 28
"...frequent changes" etc, see *Liquid Modernity*, Zygmunt Bauman (London, Polity: 2000).

"Proleptic elegy", see *Dark Vanishings*, Patrick Brantlinger (Cornell University Press, 2003).

A Billion Black Anthropocenes or None, Kathryn Yusoff (University of Minnesota Press, 2018).

Wounded Healers, Henri Nouwen (Crown, 2013).

Chapter 29
First written for <www.christianclimateaction.org>.

Chapter 30
Compiled from interviews with Jeremy Williams and Louise Williams, October 2019.

Chapter 32
Giles Goddard interviewed by Jeremy Williams, November 2019.

Additional quotes from St John's Waterloo blog, <https://stjohns waterloo.org/blog/44446> and 'Extinction Rebellion thanks churches for their support', Ellen Teague and Liz Dodd in *The Tablet*, 31st October 2019. https://www.thetablet.co.uk/news/12175/extinction-rebellion-thanks-churches-for-their-support->

Chapter 33
Stories compiled from interviews with Jeremy Williams and a survey on <christianclimateaction.org>. Additional quotes from Things Unseen podcast, Arrested to Save the Earth <https://www.thingsunseen.co.uk/podcasts/arrested-to-save-the-earth/>.

Additional material from 'Rebel Reverend Mark Coleman arrested during Extinction Rebellion protest', *Rochdale Online*, 09 October 2019.

NLT – verse taken from the New Living Translation.

Chapter 34

Compiled from a survey for Christian Action Climate.

Chapter 35

Stories from interviews with Jeremy Williams and Louise Williams, and online Christian Climate Action survey.

Quotes from Elsie Luna from her speech to Extinction Rebellion, April 2019.

Chapter 36

For Gene Sharp's list, see his book *The Politics of Nonviolent Action, Vol 2: The Methods of Nonviolent Action* (Sargent, 1980). It is also available online.

Stories drawn from interviews with Jeremy Williams, and a survey run on the Christian Climate Action website.

David Benjamin Blower quote on revelation from 'Red Rebels and the Apocalypse', Soil Journal. <https://us20. campaign-archive.com/?u=e17f405d3d804e83fb6893954&id=a99fc7d296>.

Margaret Bullit-Jonas quote from her sermon for the Convention Eucharist, Episcopal Diocese of Western Massachusetts, held at Tower Square Hotel, Springfield, MA, November 9, 2019.

<http://revivingcreation.org/a-sacramental-life-rising-up-to-take-climate-action/>.

Chapter 37

South Africa, Green Anglicans: <http://www.greenanglicans.org/

the-anglican-church-of-southern-africa-declares-a-climate-emergency/>.

Pope Francis <https://climateemergencydeclaration.org/leader-of-1-3-billion-catholics-declares-a-climate-emergency/>.

Diocese of Guildford: <https://www.cofeguildford.org.uk/whats-on/news/detail/2019/08/12/surrey-bishops-declare-climate-emergency-and-pledge-7-climate-care-commitments]>.

Methodist Church <https://www.methodist.org.uk/about-us/news/ latest-news/all-news/methodist-church-recognises-a-climate-emergency/>.

Chapter 38

Grief circles inspired by the work of Joanna Macy.

Chapter 39

Confession and lament – Jon Swales, first used at St George's Leeds.

ESV – verses taken from the English Standard Version.

Litany for the Earth – Fran Pratt, used with permission.

NLT – verses taken from the New Living Translation.

Litany of resistance – Jim Loney, adapted with permission. Originally written for Christian Peacemaker Teams.

A Prayer for the Climate, reproduced with thanks to Christian Aid.

Solemn Intention Statement reproduced courtesy of Extinction Rebellion.

Image credits